Zero-Energy Buildings - New Approaches and Technologies

Edited by Jesús Alberto Pulido Arcas, Carlos Rubio-Bellido, Alexis Pérez-Fargallo and Ivan Oropeza-Perez

Published in London, United Kingdom

IntechOpen

Supporting open minds since 2005

Zero-Energy Buildings - New Approaches and Technologies
http://dx.doi.org/10.5772/intechopen.87727
Edited by Jesús Alberto Pulido Arcas, Carlos Rubio-Bellido, Alexis Pérez-Fargallo and Ivan Oropeza-Perez

Contributors
Modeste Kameni Nematchoua, Sigrid Reiter, Aissa Bouaissi, Long Yuan Li, Mohd Mustafa Al Bakri Abdullah, Romisuhani Ahmad, Rafiza Abdul Razak, Zarina Yahya, Rim Zgueb, Amal Brichni, Noureddine Yacoubi, Dirk Schwede, Auwalu Faisal Koko, Muhammed Bello, Collins Nwaokocha, Abayomi Layeni, Samuel Tongo, Oluwasegun Alabi, Kasali Aderinmoye Adedeji, Olusegun Samuel, Taiwo Samuel, Jacob Adebayo, Olabode Olanipekun, Solomon Giwa, Olalekan Olamide, Olawale Onabanjo, Folashade Oniyide, Jagun Zaid Oluwadurotimi, Olaolu Folorunsho, Maher Shehadi, Mostafa Esmaeili Shayan

First published in London, United Kingdom, 2021 by IntechOpen
IntechOpen is the global imprint of INTECHOPEN LIMITED, registered in England and Wales, registration number: 11086078, 5 Princes Gate Court, London, SW7 2QJ, United Kingdom
Printed in Croatia

British Library Cataloguing-in-Publication Data
A catalogue record for this book is available from the British Library

Additional hard and PDF copies can be obtained from orders@intechopen.com

Zero-Energy Buildings - New Approaches and Technologies
Edited by Jesús Alberto Pulido Arcas, Carlos Rubio-Bellido, Alexis Pérez-Fargallo and Ivan Oropeza-Perez
p. cm.
Print ISBN 978-1-78985-245-5
Online ISBN 978-1-78985-246-2
eBook (PDF) ISBN 978-1-78985-355-1

For EU product safety concerns:
IN TECH d.o.o., Prolaz Marije Krucifikse Kozulić 3, 51000 Rijeka, Croatia, info@intechopen.com or visit our website at intechopen.com.

Meet the editors

Jesús Alberto Pulido Arcas graduated in architecture from the University of Sevilla (Spain), where he also obtained his M.Sc. (2009) and his Ph.D. (2013). He has extensive work experience as a professor and researcher in Spain (University of Sevilla), Chile (The University of The Bio-Bio), and now he works as a Project Assistant Professor at The University of Tokyo (Japan). His area of expertise covers heat transfer in buildings, CFD, energy efficiency, adaptive comfort, BIM technologies, and climate change in the building industry. He is the author of more than 30 research outputs in international peer-reviewed journals and a recent book on energy efficiency in buildings.

Prof. Carlos Rubio-Bellido is a Professor in the Department of Building Construction II at the University of Seville. His research is focused on energy efficiency in the building sector as well as building performance simulation. He is a Visiting Professor at Bío-Bío University (Chile). He is a member of the International Scientific Committee of various international conferences. He is the author of more than 50 research papers. He is a recognized reviewer of various international indexed journals and international research projects.

Ph.D. Alexis Pérez Fargallo Graduated in architecture from the University of Sevilla (Spain). He obtained his MSc (2012) and his PhD (2013) from the University of Seville, Spain. He won a postdoctoral fellowship from the Conicyt (Chile) and carried out his postdoctoral research at the University of Bío-Bío. He later started working as an Assistant Professor in the Department of Building Science at the University of Bio-Bio, Chile. His area of expertise covers energy renovation, energy costs, energy efficiency, thermal comfort, and fuel poverty. He is an author of more than 30 research outputs in international peer-reviewed journals and recently authored a book on energy efficiency, and has directed and participated in several research projects. He has carried out an advisory, direction, design, production, and construction roles in architectural projects. He has given lectures on energy renovation and fuel poverty in several countries.

Ivan Oropeza-Perez has his expertise in thermal comfort, passive cooling systems, and sustainable architecture. The main purpose of his research is to achieve a proper indoor environment while saving energy and water. He has written a book regarding natural ventilation within buildings and several research articles with similar topics. He has participated in dozens of international conferences around the world presenting works on the aforementioned subjects. Currently, he is an Associate Professor Senior in the Architecture Department of the Universidad de las Americas Puebla in Mexico. He is also the President of the International Building Performance Simulation Association section Mexico, a fellow of the Mexican National Solar Energy Association, and a fellow of the Mexican National Researchers System of the Mexican National Council of Science and Technology.

Contents

Preface

The building industry has been undergoing significant changes in the last decades. Traditionally labeled as unproductive and inefficient, it is evolving into a more environmentally conscious economic sector, and energy efficiency is one of the main vectors of this transformation. In this context, Zero-Energy Buildings were born as a technological paradigm, a building that virtually uses no energy, through a combination of advanced insulation systems and locally produced renewable energy. This concept has been exported to many countries that are facing the challenge of reducing the energy consumption of their building stock, although not without controversy; its adaptation to different social, economic, and cultural contexts is not always a one-way process. This book aims at providing a broad view of the variegated facets of zero-energy buildings by compiling experiences from scholars with different backgrounds.

The book is organized into three main sections. Section 1 is the implementation of zero-energy buildings in different contexts, in which contributing authors shed light on the numerous challenges that these technologies face when being adapted to local contexts. Section 2 is on renewable energy, which provides insights into the integration of solar radiation and natural ventilation into the design process. Section 3 is sustainable materials, which presents research focused on the improvement of the thermophysical properties of cement and concrete, with the ultimate aim of improving the insulation levels of the thermal envelope.

It is expected that this book will be of use to scholars, researchers, students, and practitioners in the field of energy-efficient architectural design whose particular interest pertains to zero-energy buildings. The variegated contributions assure a plurality of perspectives that aims at enriching its conception, not so much as an unmuted product exported to foreign contexts, but as a flexible concept that may be impregnated with different technologies and be able to accommodate the particularities of countries and people in terms of their social and cultural context.

This book has been possible thanks to the significant contribution of the authors, to whom the editors would like to express their sincere gratitude for their effort, especially during these uncertain times that have put on hold many of our research achievements. Lastly, the editor would like to thank the relentless effort and continuous support of their colleagues. Despite being in distant corners of the world, the millenary city of Cadiz in Spain, the always green and lively city of Concepción in Chile, and Tokyo in Japan, the city that never stops amazing, they have managed to complete this book, which will be one of many projects to come in the future.

Jesús Alberto Pulido Arcas, Ph.D.
Project Assistant Professor,
The University of Tokyo,
Japan

Carlos Rubio Bellido, Ph.D.
Professor in the Department of Building Construction II,
The University of Sevilla,
Spain

Alexis Pérez Fargallo, Ph.D.
Assistant Professor in the Department of Building Science,
The University of The Bio-Bio,
Chile

Ivan Oropeza-Perez
Associate Professor Senior in the Architecture Department,
Universidad de las Américas Puebla,
Mexico

Section 1

Implementation of Zero-Energy Buildings in Different Contexts

Chapter 1

Net Zero Energy Buildings and Low Carbon Emission, a Case of Study of Madagascar Island

Modeste Kameni Nematchoua and Sigrid Reiter

Abstract

The buildings respecting the concept "Net Zero energy" are becoming more and more flowering in the world these last years. The main goal of this research is to evaluate the different possibilities of implementation of buildings with Net zero energy and low environmental impacts in Sub-Saharan Africa. The proposed building is 80% made of local materials with low carbon emissions and especially at lower cost. The optimization and modeling of the building is carried out by the Design Builder software, which is a world-renowned software in the field of optimization of comfort, cost, carbon reduction, etc. By fixing the insulation thickness up to 11 cm, cooling and heating energy are found equal to zero during the different operating seasons in this residential building. The results show that the optimal solution to consider a net zero energy building in Antananarivo city requires an additional expense estimated at 40% of the cost of buildings more conventional encountered in the island. This will save \$475 each year starting in 2030, with 99% reduction in the CO_2 release. The choice of local materials with low conductivity, low emissions, and low cost, has a significant impact on the implementation of a sustainable building, and more adapted to climate change concept.

Keywords: Net Zero energy, low carbon, building, island

1. Introduction

Nowadays, climate change has become one of the major concerns of all governments and politicians around the world [1, 2]. According to the IPCC, it is expected a temperature increase ranging from 2 to 3.5°C depending on the region [3, 4]. The results proved by some experts in this field showed that sub-Saharan Africa was one of the most vulnerable regions [5, 6]. The objectives of the European Union oriented towards the energy efficiency stipulate that the design of the ecological buildings and more adapted to the new current climate can be a solution to this plague [7]. The search for carbon neutrality by trying to live in the most ecological way is a good thing, but living, in addition, in "green" buildings is even better [8]. The notion of sustainable building varies according to the scientific and specialist fields. Overall, it is a healthy construction, using natural materials [7]. A building must first of all adapt to the man, the well-being of the occupants being capital [9]. Specialists in this field condemn the use of toxic substances in the industrial manufacture of building materials. The key role of energy conservation specialists is to limit the negative impacts of human habitat on the environment, using state-of-the-art technologies, and to reduce the energy consumption of buildings, houses and buildings [10]. Indeed, they advocate reinforced thermal insulation and sharp construction techniques. "Eco-Builders" consider the building throughout its life. In addition to saving energy, they are also concerned about the origin of the materials used and their management (disposal, recovery) at the end of their life. Sustainable construction is not a specific construction method, but it brings together a set of techniques, materials and technologies that, properly integrated into a building, contributes to enhancing its environmental performance [11]. In its ideal embodiment, green building optimizes energy efficiency, limits water consumption, makes maximum use of recycled, recyclable and non-toxic materials and generates the least amount of waste possible during construction as well as occupation [12]. Around 40% of total energy use and around 24% of CO_2 emissions come from worldwide energy use in buildings. Energy use and emissions include both direct, on-site use of fossil fuels as well as indirect use from electricity, district heating/cooling systems and embodied energy in construction materials [13]. The term ZEB is commonly used as propaganda without any mastery of the different realities and constraints specific to each country. There are several methods and strategies to reach net-zero energy, however, for NetZEBs to become mainstream in the international market, it requires several consensus on clear definitions, and also agreement on the building performance which could inform "zero energy" building policy, programs and industry building practices, as well as design tools, case studies and demonstration projects that would support industry adoption [14]. The different objectives of Net zero energy were considered to be a new proposal that was inaccessible because of its high cost, they were recommended only for large projects. Today, it is now possible for all types of construction [15]. Net Zero-Energy Building has become a popular catch phrase to describe the synergy between energy-efficient building and renewable energy utilization to achieve a balanced energy budget over an annual cycle. Several experts have proposed different methods for designing zero energy and carbon buildings in several types of climate [16–22]. But no study has yet been done in the specific case of a country in the Indian Ocean (Madagascar, Mayotte, Reunion, Comoros etc.). This study has for fundamental objective to propose a model of implementation of buildings to Net Zero energy in Madagascar. Section 2 gives the different stages of implementation of this district.

2. Methodology

2.1. Studied place

Antananarivo is a city overlooking a hill that rises 1248 meters above an area of rice paddies. It is the capital of Madagascar and it is also the economic and administrative lung of this country. Antananarivo has a tropical climate of altitude. Although it is located in the inter-tropical zone, the average temperature over the year is moderated by the effects of altitude. The climate is characterized by cool and very dry winters and mild and very wet summers. Cool season rarely drops below 10°C. In the hot season, it rarely exceeds 32°C. The choice of this city for the implementation of this type of building is not random, indeed, the climate of this city is very favorable to the implementation of eco buildings. The building materials used in the architecture of this building is cheap in this city. The deficits and power cuts are regularly observed in this city. The popularization of this building throughout the country can help the government to solve this problem.

2.2. Climate parameters

The different outdoor weather data of the last 15 years of Antananarivo are downloaded with the Meteonorm software and applied to this study. The Meteonorm software is one of the world's best meteorological software that allows you to connect in a few minutes to a meteorological station of the city you are looking for or to the nearest city and to download the climatic data specific to this city. The output data can be recorded on several types of freely selected formats.

2.3. Comfort parameters

The different comfort parameters are evaluated in this study with the Design Builder software. It is possible to calculate the operating temperature of the building (uniform temperature). We evaluated the PMV (Predicted Mean Votes) of Fanger, Kansas and Pierce. Comfort hours are assessed based on ASHRAE recommendations with 90% occupant acceptability, CEN15251 Category I, II and III.

2.4. Description of building

The building evaluated is a family house spread over an area of 273 m2, and consists of: three bedrooms, shower, living room and kitchen, occupied by 5 people. The modeling of the building is shown in **Figure 1**.

Table 1 showed the different characteristics of materials applied in this building.

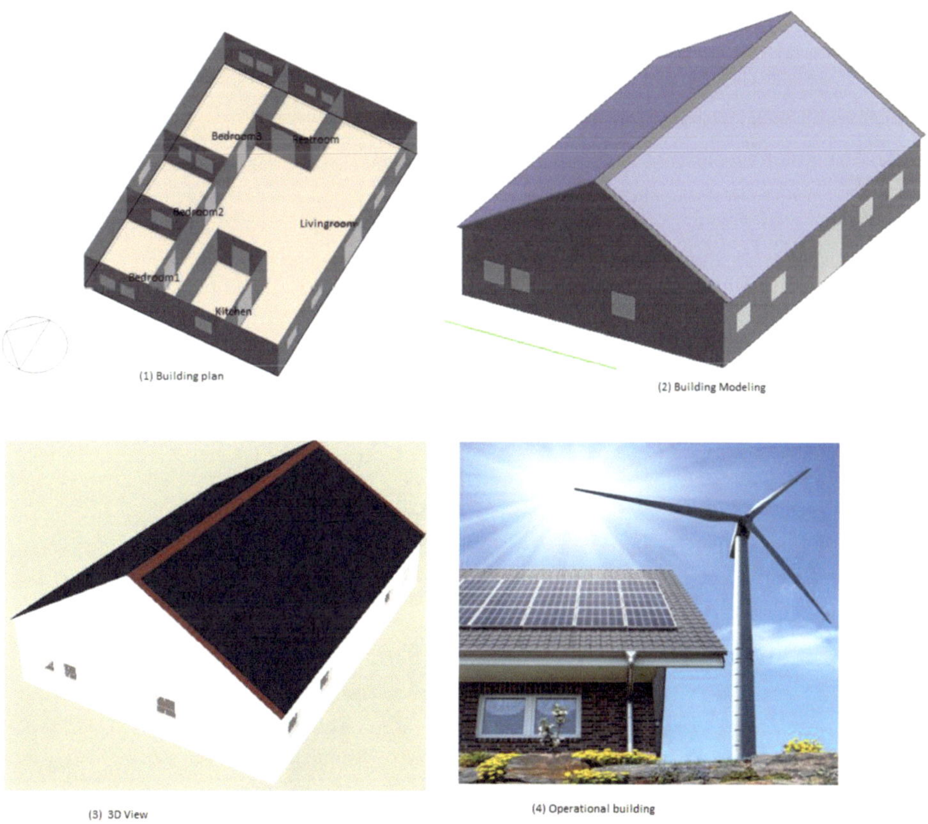

Figure 1.
Studied building.

	Layer	Component	Thickness (m)	Thermal conductivity (W/m-K)	Density (kg/m3)	Specific heat capacity (J/kg K)	Embodied Carbon (kgCO2)	Cost ($/m2)	U-value (W/m2-K)
Building element	Layer1	Ceiling	0.02	0.056	380.0	1000.0	1.2	6.6	0.20
	Layer2	Hemp	0.09	0.04	25.0	1000.0	0.0	27.5	
	Layer3	Limestone silicon	0.10	0.136	270.0	880.0	0.0	16.5	
	Layer4	wood	0.05	0.056	380.0	1000.0	1.2	6.6	
	Layer1	Roof tiles	0.030	0.08	530.0	1800.0	0.19	30	

Table 1.
Thermal properties of some materials.

It was seen in this table that the materials as Hemp and Limestone Silicon do not produce CO_2.

2.5. Description of software

The modeling of the building and all simulations were led thanks to Design Builder. The Design Builder software is one of the most famous existing software in modeling and optimization of the building. It also helps reduce the carbon content. The most recent version 6.3 is used in this study. The Design Builder tool also minimizes costs and hours of discomfort.

2.6. Calibration of model

To calibrate this model, we compared the different simulated and measured values of a building typically encountered in Madagascar [23], by calculating the linear correlation coefficient (R^2) to analyze the margin of error. The literature shows that the error is negligible if the correlation coefficient obtained is around of ±1.

2.7. Wind turbine and photovoltaic systems

Wind turbine with alternating current worked 24/7. This wind turbine was a rotor type horizontal with a diameter estimated to 41 m, a height of 31 m, number of blades 3, with a maximum power coefficient of 0.4.

The photovoltaic panels occupied almost three-quarters of the roof area, making an angle of 45°C, with maximum orientation from south to north. The different panels consisted of polycrystalline cells with a mixed association.

2.8. Scenarios

In the reference scenario, we decided to study this residential building without any physical constraint. In its state as naturally as possible, and there is no source of electrical production. In this case, we use the A2 scenario, designed by the IPCC, which is the most realistic in Madagascar [24] for assessing indoor air quality and temperature in the future.

In a scenario 1, we install photovoltaic panels on two-thirds of the roof, while increasing the thickness of insulation by 2 cm (from 9 to 11 cm). The main facade is oriented from south to north, with solar protection on each window. The inclination of the solar panels is set at 45°C. The network was not connected to a power storage system (e.g., the battery).

In Scenario 2, we apply all the details presented in Scenario 1, except that the entire power grid is connected to a storage system. In addition to this, we apply the wind turbine to the building, whose characteristics are detailed in the previous paragraph. We made simulation this building according to each scenario and we got found new results.

3. Results and discussions

3.1. Indoor air

Air temperature and relative humidity are both environmental parameters which their variation has a significant impact on the occupant's comfort. **Figure 2** shows the variation of indoor air temperature in the new building. We can see that currently, in the building, indoor air temperature varies from 19.83 to 22.57°C; in 2030, the indoor air temperature is expected to be between 19.96 and 22.82°C; in 2050, in the same condition, indoor air temperature will be between 20.41 and 23.10°C. Globally, indoor air will increase to 0.30°C in 2030, and 0.52°C in 2050; compare to current air temperature. In addition, it is seen in **Figure 3** that presently, relative humidity varies between 59.57 and 75.41%. In the future, it will vary between 58.77 and 76.03% in 2030, and, from 59.82 and 77.25% in 2050. The analysis showed that relative humidity will increase to 1.51% in 2030, and 2.73% in 2050; compare to 2017. The ASHRAE 55 ranges of comfort suggested indoor air temperature of 23–26°C; and relative humidity of 30–60% [25]. These different values are outside the ASHRAE ranges. Antananarivo is dominated by several mountains which affect the climate of this city. This interval is low compare to that found by Nematchoua et al. [26], in traditional buildings in Madagascar, which were between 24.6 and 28.4°C.

3.2. Electricity

Figure 4 analyzed the potential of total energy demand and produce by this building. Monthly electricity consumption varies between 123.8 and 137.1 kWh.

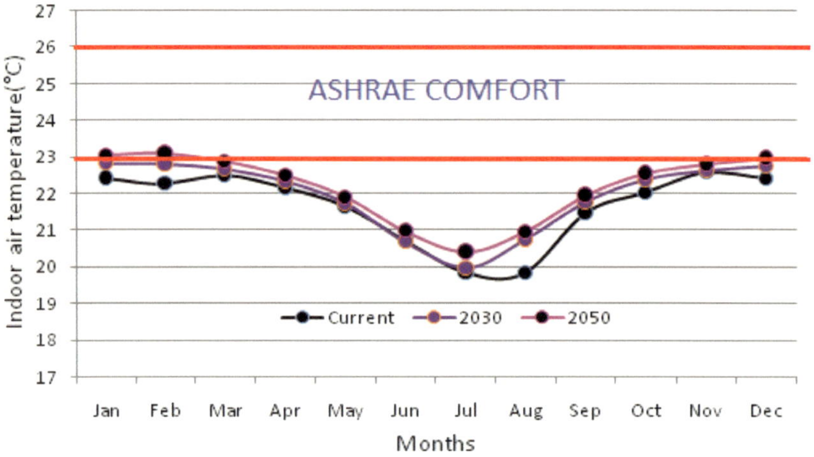

Figure 2.
Monthly indoor air temperature in the new building distributed on three periods (current, 2030 and 2050).

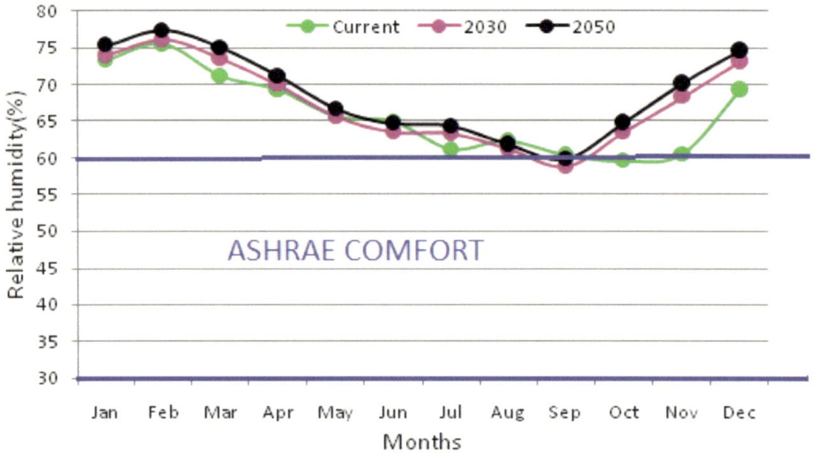

Figure 3.
Monthly relative humidity in the new building during three periods (present, 2030 and 2050).

We can see under basic of this scenario electricity generated by this building corresponds net to electricity consumption; with zero cooling energy building during the different seasons. In the specific case of this scenario, which simply recommends an application of solar panels covering a total area of 182 m^2, Net Zero Energy Building objectives are achieved for this building(energy produce = energy consumption). Electricity generated was estimated to be around of 0.49 kWh/m^2. In the second scenario which some results are showed on **Figure 5**, it was applied simultaneously wind turbine and photovoltaic panel on the building.

It is noticed that in this case, the electricity generated by the building is equal to 13 times the average electricity consumed by the building. At this precise moment of operation of the building, the new building can be considered as a building with positive energy, that is to say it produces more than it consumes (energy consumption < energy produced). The annual total electricity that can be sold to individual consumers is estimated to 18946.86 kWh per year; it allows to save 4550$. The frequency of comfort and total energy consumption is showed on **Figure 6**.

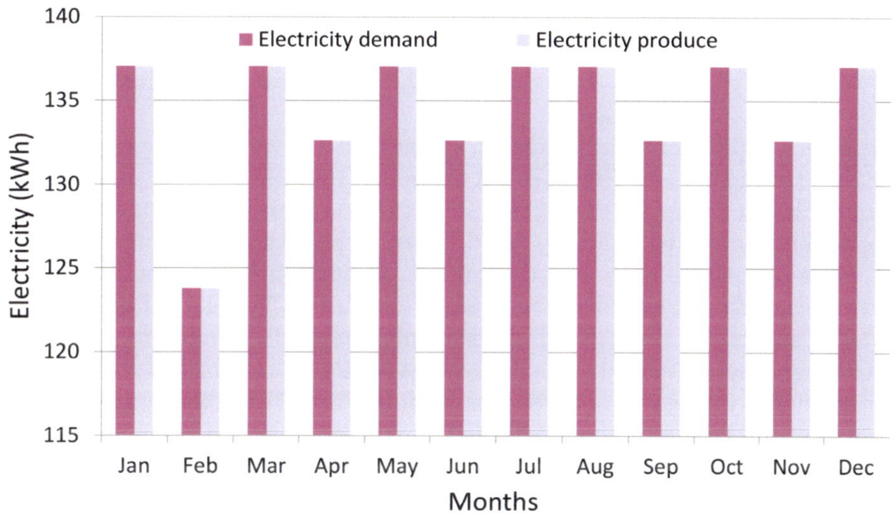

Figure 4.
Scenario of Net Zero Energy of building.

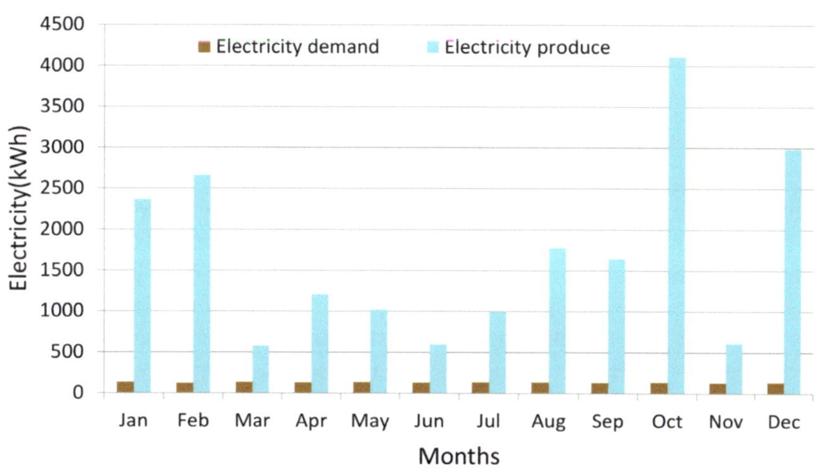

Figure 5.
Application of wind turbine and photovoltaic panel on the building.

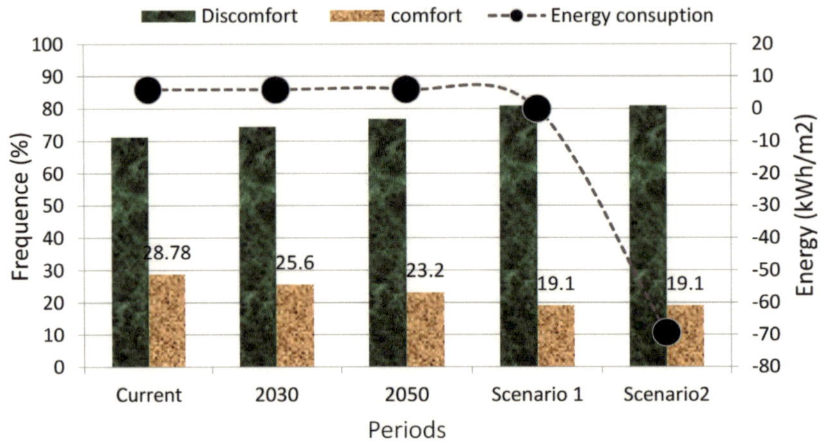

Figure 6.
Comfort potential and building energy consumption.

Discomfort potential was estimated to 71.2% (current); 74.4% (2030), and 76.8% (2050). These results show that in 2050, indoor air of the building will be 5.6% more uncomfortable than currently.

It is very important to notice that in specific case of scenario1, energy demand is found at zero. This does not mean that the building does not consume energy, it is just to explain that at this point the energy production is equal to the consumption of the building. The different energy values assigned a sign (−), explain that at this moment there is overproduction. These results are very interesting, and can be used by the building specialists. The electrification rate in Madagascar is one of the lowest in Africa: only 15% of the inhabitants are connected to an electricity grid. This figure rises to 58% in urban areas and drops to 4.7% in rural areas, which nevertheless accounts for 70% of the country's population. It would be recommended to the Malagasy government to create favorable conditions to encourage the population to design new buildings more ecological and comfortable. One of the limitations of this research is that the type of building proposed costs up to 40% more expensive than the more conventional buildings found in the big island. But today, it is revealed in the literature that only 2% of the Malagasy population would be able to build this kind of building. We are well aware of this, but we think that the ideal for a more sustainable solution is to build new buildings in the big island by respecting the criteria mentioned in this study.

4. Conclusion

In this research, we analyzed and suggested a model allowing to reach net zero energy building and in certain measure created a building with positive energy in Antananarivo. Operational carbon was estimated to be around 3.7 $kgCO_2/m^2$. The operative temperature was between 19 and 23°C, in this period, the comfort potential was from 30%. The results found in this study showed it is possible to reach objective "Net Zero energy building" in Madagascar island by respecting the way detailed in this research. The degree of vulnerability in climate change is very high in Madagascar. The Malagasy government should propose more reliable control and adaptation strategies, for example the case of the extension of ecological buildings is very interesting. In a future study, we will study the case of implementation of the concept Net zero energy neighborhood.

Author details

Modeste Kameni Nematchoua[1,2*] and Sigrid Reiter[2]

1 Beneficiary of an AXA Research Fund Postdoctoral Grant, Research Leaders Fellowships, Paris, France

2 Management and Analysis (LEMA), Liège, Belgium

*Address all correspondence to: kameni.modeste@yahoo.fr; mkameni@uliege.be

References

[1] Nematchoua MK, Orosa JA, Reiter S. Climate change: Variabilities, vulnerabilities and adaptation analysis—A case of seven cities located in seven countries of Central Africa. Urban Climate. 2019;**29**:100486

[2] Nematchoua MK, Ricciardi P, Orosa JA, Buratti C. A detailed study of climate change and some vulnerabilities in Indian Ocean: A case of Madagascar Island. Sustainable Cities and Society. 2018;**41**:886-898

[3] IPCC et al. The physical science basis. In: Solomon S, Qin D, Manning M, Chen Z, Marquis M, Averyt KB, editors. Contribution of Working Group I to the Fourth Assessment Report of the Intergovernmental Panel on Climate Change. Cambridge: Cambridge University Press; 2007. pp. 1-180

[4] IPCC. Climate Change 2001: The Scientific Basis. Contribution of Working Group 1 to the Third Assessment Report of the Intergovernmental Panel on Climate Change. Cambridge, UK/New York, US: Cambridge University Press; 2001. pp. 25-39

[5] IPCC. Climate Change: The Physical Science Basis. Contribution of Working Group I to the Fourth Assessment Report of the Intergovernmental Panel on Climate Change. UK: Cambridge University Press; 2007. pp. 1-21

[6] IPCC. Climate Change: Impact, Adaptation and Vulnerability. Summary for Policymakers, Working Group II Contribution to the Fifth Assessment Report of the Climate. UK: Cambridge University Press; 2014. pp. 1-34

[7] Comité Européen de Normalisation. En 15978. Contribution des Ouvrages de Construction au Développement Durable—Évaluation de la Performance Environnementale des Bâtiments—Méthode de Calcul. Bruxelles, Belgique: Comité Européen de Normalisation; 2011

[8] International Energy Agency Transition to Sustainable Buildings. Strategies and Opportunities to 2050. 2013. ISBN: 978-92-64-20241-2

[9] Nematchoua MK, Tchinda R, Orosa JA. Thermal comfort and energy consumption in modern versus traditional buildings in Cameroon: A questionnaire-based statistical study. Applied Energy. 2014;**114**:687-699

[10] EPBD Recast: Directive 2010/31/EU of the European Parliament and of the Council of 19 May 2010 on the Energy Performance of Buildings (Recast). Available from: http://ec.europa.eu/energy/efficiency/buildings/buildings_en.htm; http://eur-lex.europa.eu/johtml.do?uri=oj:l:2010:153:som:en:html

[11] Sartori I, Napolitano A, Voss K. Net zero energy buildings: A consistent definition framework. Energy and Buildings. 2012;**48**:220-232

[12] ASHRAE Vision 2020: Producing Net Zero Energy Buildings. A Report from the American Society of Heating, Refrigeration and Air-Conditioning Engineers. American Press. January 2008;**1**(5)

[13] SHC Task 40—Towards Net Zero Energy Solar Buildings. SHC Position Paper. 2015. pp. 2-10

[14] McGraw Hill Construction. World Green Building Trends: Business Benefits Driving New and Retrofit Market Opportunities in 60 Countries. SmartMarket Report. 2013;**5**(10)

[15] Carmicahel C, Managan K. Reinventing Existing Buildings: Eight Steps to Net Zero Energy. Rocky

Mountain Institute, Johnson Controls; 2013

[16] Salem R, Bahadori-Jahromi A, Mylona A, Godfrey P, Cook D. Investigating the potential impact of energy-efficient measures for retrofitting existing UK hotels to reach the nearly zero energy building (nZEB) standard. Energy Efficiency. 2019;**12**(6):1577-1594

[17] Madad A, Gharagozlou A, Majedi H, Monavari SM. A quantitative representation of the urban green building model, focusing on local climatic factors by utilizing monetary valuation. Ecological Economics. 2019;**161**:61-72

[18] Wang D, Pang X, Wang W, Qi Z, Li J, Luo D. Assessment of the potential of high-performance buildings to achieve zero energy: A case study. Applied Sciences. 2019;**9**(4):775

[19] Tronchin L, Manfren M, Nastasi B. Energy efficiency, demand side management and energy storage technologies—A critical analysis of possible paths of integration in the built environment. Renewable and Sustainable Energy Reviews. 2018;**95**:341-353

[20] Fořt J, Beran P, Pavlík Z, Černý R. Complex assessment of reconstruction works on an institutional building: A case study. Journal of Cleaner Production. 2018;**202**:871-882

[21] Thomas A, Menassa CC, Kamat VR. A systems simulation framework to realize net-zero building energy retrofits. Sustainable Cities and Society. 2018;**41**:405-420

[22] Aghamolaei R, Shamsi MH, Tahsildoost M, O'Donnell J. Review of district-scale energy performance analysis: Outlooks towards holistic urban frameworks. Sustainable Cities and Society. 2018;**41**:252-264

[23] Nematchoua MK, Ricciardi P, Buratti C. Statistical analysis of indoor parameters an subjective responses of building occupants in a hot region of Indian ocean; a case of Madagascar island. Applied Energy. 2017;**208**:1562-1575

[24] Nematchoua MK, Yvon A, Kalameu O, Asadi S, Choudhary R, Reiter S. Impact of climate change on demands for heating and cooling energy in hospitals: An in-depth case study of six islands located in the Indian Ocean region. Sustainable Cities and Society. 2019;**44**:629-645

[25] ASHRAE. ANSI/ASHRAE Standard 55R, 'Thermal Environmental Conditions for Hum Occupancy'. Atlanta: American Society of Heating, Refrigerating and Air-Condition Engineers Inc.; 2004

[26] Nematchoua MK, Ricciardi P, Buratti C. Adaptive approach of thermal comfort and correlation between experimental data and mathematical model in some schools and traditional buildings of Madagascar under natural ventilation. Sustainable Cities and Society. 2018;**41**:666-678

Chapter 2

Exploring the Factors Hindering the Use of Green Architecture in Nigeria

Auwalu Faisal Koko and Muhammed Bello

Abstract

The construction industry in Nigeria has continuously witnessed rapid development as a result of massive investments in infrastructural projects such as housing. The continuous growth of this industry and the conventional approach to construction practices in Nigeria have negatively affected the environment and the wellbeing of the populace. Therefore, the concept of green architecture, also known as sustainable architecture, is a new approach in Nigeria's construction industry that strives to achieve environmental sustainability. However, various factors have hindered its adoption and utilisation. This study, therefore, examined the various factors hindering the use of green architecture through various literature reviewed and administered questionnaires to built environment professionals in Nigeria to ascertain their perception of those identified factors. Data gathered from the questionnaires were analysed using descriptive statistical tools and ranked according to each factor's mean index score and relative importance index. The results of the study revealed the most prominent factors hindering the utilisation of green architecture in Nigeria. Hence, findings from this study suggest that more efforts such as public enlightenment and the provisions of incentives are needed to be done by the government, built environment professionals, and other stakeholders in Nigeria's construction industry for the promotion of green architecture.

Keywords: built environment, green architecture, green buildings, sustainable buildings, zero-energy

1. Introduction

The term 'Green Architecture' and 'Sustainable Architecture' are often used interchangeably. Green Architecture refers to an architecture that is environmentally friendly and resource-efficient throughout the building life cycle from siting to design, construction, operation, maintenance, renovation, and demolition [1, 2]. Green Architecture practice complements the current building concerns of economy, utility, durability, and comfort. Even though new technologies are continuously being developed to improve the current practices in creating a sustainable built environment. The practice of green architecture allows for the improvement of the general living standard without causing damage to the resources needed for our survival as humans [3]. Green architecture helps

IntechOpen

conserve natural resources, reduce pollution, and prevent environmental degradation [4]. The primary aim of green architecture is to reduce the overall impact of the built environment on human health and the natural environment. Green architecture principles often lead to the lowering of the operational cost of buildings through reduced utility from energy and water use. It also leads to a reduction in building maintenance costs. The green approach to architecture has been in existence for years, and it is not a new approach in trying to reduce the environmental impact of architecture. According to Brenda and Robert, what seems to be new is the recognition that green approach to natural and built environment involves a holistic approach to the design of buildings; with consideration to all the resources that go into a building, such as materials, fuels and the contribution of the users [5]. In Nigeria, the concept of green architecture, its principles, as well as its advantages to the environment, is hardly put into consideration when designing new buildings or renovating buildings. The result presented by Otegbulu revealed inadequate satisfaction of building users, which is often as a result of neglected green building principles during design, construction, and maintenance phases of building construction [6]. Therefore, Green Architecture seeks to meet the needs of humans for food, shelter, natural resource, transportation, and effective waste management while preserving and protecting the quality of the environment as well as the natural resource base, which is essential for future life and development [7, 8]. This concept recognises that meeting long-term human needs will be impossible unless the earth's natural, physical, and the chemical system is conserved [9], which is in line with the concept of green architecture. There is no doubt that the principles of green architecture are important for the construction industry in Nigeria, particularly the real estate sector, which has grown steadily and rapidly in the last two decades. Figures indicate that the real estate sector as a major component of the construction industry witnessed one of the highest growth rates and contributed tremendously to Nigeria's national income in 2018 with more than ₦1.26 trillion [10]. As such, the adoption of the principles of green architecture will not only enhance but also protect communities in Nigeria. This study, therefore, seeks to explore the concept, principles, and factors hindering the use of green architecture in Nigeria.

2. Concept of green architecture

Green architecture, also referred to as sustainable architecture, is a philosophy of designing buildings to comply with the principles of social, economic, and ecological sustainability. Green architecture uses a conscious approach to energy and ecological conservation in the design of the built environment [11]. It approaches building construction from the conceptualisation at the design and construction stage to the material usage at the finishing stage and throughout the entire lifespan of the buildings in order to minimise the harmful effect on human health and the environment [12]. Elshimy defined green architecture as the design and construction practices that considerably reduce or eliminate the undesirable effect of buildings on the environment and its occupants [13]. Similarly, the Integrated Waste Management Board, as cited in Kadiri, defined green architecture as an architecture that is designed, constructed, renovated, operated, or used in an ecological and resource-efficient manner [14]. It involves a trans-disciplinary approach to building construction and takes into consideration the environment as a vital factor in the design process [15]. Therefore, the concept of green architecture strives to minimise the negative environmental impact of buildings by enhancing efficiency and modernisation in the use of materials and energy, as well as conservation of the environment [16–20]. However, despite the several benefits of green architecture, the conventional architectural practice in Nigeria often overlooks the interrelationships between a building, its materials and components usage, and the environment. This makes the buildings consume more resources causing an undesirable effect on the environment and creating a tremendous amount of waste. It subsequently results in buildings that are expensive to operate in terms of energy and water consumption and also contribute to buildings having poor indoor air quality. Therefore, it is of paramount importance that various stakeholders of the construction industry adopt the principles and practice of green architecture as a possible way of creating an environment that is friendly with resource-efficient buildings and reducing the operation and maintenance costs associated with built environment design in Nigeria.

It is against this background and the possible way of encouraging the use of green architecture in Nigeria that this study explores the various factors hindering its usage and provides possible measures to overcome these challenges.

2.1 Factors hindering the use of green architecture

Despite the numerous advantages of green architecture practice as it is related to sustainable site design, water quality and conservation, energy and environment, indoor air quality as well as material and resource conservation. The practice of green architecture is still faced with various factors that hinder its acceptance and usage. Choi et al. and Issa et al. identifies the various factors hindering the use of green architecture to include capacity barriers, the high initial cost of construction, and lack of proper awareness and enlightenment [21, 22]. Hamidi believes that the lack of green architecture expertise to initiate its concepts and principles from the early stages of building design and planning hinders the utilisation of green architecture [23]. Another factor identified by Issa et al., which hinders the development of green architecture, is the slow recovery of long-term costs [22]. Also, Means identified the various factors which hinder the use of green architecture to include low priority for the sustainability agenda in the education system, lack of urgency surrounding the practice of sustainability; lack of enforcement by the relevant authorities, and lack of government intervention through policies and incentive [24]. Other factors include financial constraints, the belief that sustainable buildings are economically not viable as they add to project costs, the belief that sustainable construction is an academic pursuit and not viable in practice, and most important is the lack of political will. In summary, the various factors hindering the use of green architecture, as identified from various literature, are categorised and discussed below.

2.1.1 Technology/capacity barrier

Various studies by different researchers have identified technology and capacity barriers as one of the main factors that hinder the development of green architecture. In a study conducted by Robichaud and Anantatmula, it identified a lack of knowledge about green architecture practices, lack of training, and education as the main barriers to the implementation of green building practices in developing countries [25]. This is further stressed by Du Plessis, Hakkinen and Belloni, and Opoku and Ahmed as a barrier due to the shortage of expertise in green architecture practices and its sustainability [26–28]. Rydin et al. argued that not only are professionals supposed to be knowledgeable, but also need to form an integrated team comprising of the developer/owner, project manager, contractor, architect, services engineer, structural engineer, civil engineer, environmental engineer, landscape consultant, cost planner, and building surveyor [29]. This view is further supported by International Labour Office, who believes that the main reason for labour shortages and lack of industry skill in the construction industry is skill requirements change [30]. This is due to the introduction of green building designs, technologies, and practices, so previously satisfactory skill sets are no longer adequate. As a result of this, Opoku and Ahmed recognised the importance of building capacity as being an essential factor in the development of green architecture practices [28]. Therefore, built environment professionals need knowledge and technology that are better adapted to the natural resources to actualise the implementation of the various principles of green architecture. It is also essential for built environment professionals to adequately understand green architecture practice to be able to ensure that their decisions and actions regarding construction reduce the burden on building users and the environment.

2.1.2 Cultural and social resistance

There is a common lack of concern about green architecture and the high tendency of maintaining conventional construction practices by the various stakeholders of the construction industry in most developing countries [31]. This neglect of green architecture practice is experienced differently through various stakeholders such as the built environment professions, design approving authorities, ministries of lands and housing, as well as local development authorities. Also, the construction industry in most developing countries is dominated by contractors and developers that are not interested in green technological changes that involve risks and extra costs [27, 32]. In such cases, construction favours the use of conventional practices and discourages other alternatives like the use of green architecture construction methods. The construction industry in Nigeria favours the use of sandcrete blocks and reinforced concrete, which is professionally termed wet construction, and neglects other forms of sustainable construction practiced globally. Another factor determined by Du Plessis to be hindering the use of green architecture is the low interest in sustainable construction practice by clients and other stakeholders in the construction industry [27]. This is caused as a result of ignorance by clients as it relates to the long-term advantages of utilising green architectural practices. Therefore, in order to revolutionise the current construction practice, especially as regards to construction methods and materials used. The various professionals of the construction industry must enlighten clients, and other stakeholders on the competitive advantages of utilising green architecture practices for building construction.

2.1.3 Higher perceived cost associated with green architecture

Various studies have identified the fear of higher investment costs for sustainable buildings as compared to conventional buildings and the risks of unforeseen costs as one of the barriers to the utilisation and practice of green architecture [26, 32–34]. The added cost of incorporating green architecture features into building projects, which mainly depends on local factors such as climate, building customs, and labour skill levels, often serves as a significant barrier to having green buildings in most developing countries like Nigeria. Hydes and Creech believe that the high cost of green architecture practice is mainly due to overestimating energy efficient cost measures, increased consultancy fees, and underestimation of cost-saving measures [35]. This high cost usually discourages the practice of green architecture. Therefore, in order to promote the practice of green architecture. Shi et al. suggested the incorporation of Life Cycle Cost (LCC) in all construction projects by built environments professionals to ascertain the cost implication and competitive advantages of green architecture usage [36].

2.1.4 Lack of incentives that promotes green architecture

The construction industry in most developing countries seems to be lagging in terms of the provision of incentives to contractors who meet green building ratings and consultants who incorporate principles of green architecture into their designs [37–39]. This aligns with the assertion of Chan et al., Darko and Chan, as well as Darko et al., who identified the lack of government incentives as a critical barrier affecting the adoption of Green Building Technologies [40–42]. Therefore, in order to encourage the utilisation of sustainable architecture, the government needs to establish effective financial incentives and non-financial incentives schemes that would help to ease the high initial costs associated with green architecture. Arditi and Yasamis, Serpell et al., as well as Sodagar and Fieldson, believe that the provision of incentives can be used for promoting green buildings in construction contracts in order to reduce contract costs, minimise contract duration, and maintain an acceptable level of health and safety [43–45]. This will subsequently lead to productivity, technological progress; innovation; management efficiency; and satisfactory quality of construction. Therefore, the provision of incentives will undoubtedly reduce the high start-up cost associated with green architecture and promote its usage and development.

2.1.5 Limited knowledge and information regarding the economic benefits and prospects of green architecture

Inadequate knowledge and information regarding the economic benefits and prospects of green architecture also serve as a barrier to the utilisation of green architecture. The availability of information as regards to the competitive advantages the use of green architecture offers is considered to be worrisome in most developing countries [46–48]. William and Dair and Azeem et al. identified a lack of knowledge, understanding, and information as a significant barrier to the successful delivery of sustainable architecture [37, 49]. Similarly, Alabi observed a low level of awareness regarding the concept of sustainability among most construction stakeholders in Nigeria [50]. Therefore, for green architecture to be widely accepted and used for construction projects, there is an urgent need for public awareness regarding the financial, economic, and environmental benefits of green architecture to society. Also, the numerous advantages of green architecture must be documented and communicated to all relevant stakeholders in order to expand and its use. Furthermore, there should be platforms which help built environment professionals to quickly disseminate information regarding the benefits of green architecture to the general public.

2.1.6 Unavailability of green building materials

Another barrier to the adoption and use of green architecture is the scarcity of green products and materials in the building construction industry. Various studies highlighted that most construction projects in developing countries faced difficulties in sourcing green products locally [49, 51–53]. Environmental friendly products that impact less on the environment and are needed for the utilisation of the principles of green architecture are not easily and readily available for use in the building construction industry. Even when these products and materials are available in developing countries such as Nigeria, the delivery time is usually lengthy.

S/No	Factors hindering the use of green architecture	Sources
1.	Technology/capacity barrier	[25–28]
2.	Cultural and social resistance	[27, 31, 32, 54]
3.	High perceived cost associated with green architecture	[25, 26, 32–34, 55]
4.	Lack of incentives that promotes green architecture	[21, 37, 40, 56]
5.	Limited knowledge and awareness regarding the economic benefits and prospects of green architecture	[37, 47–50, 57]
6.	Unavailability of green building materials	[49, 51–53]
7.	Insufficient support from the government	[41, 42, 44, 45, 58]
8.	Preference for other conventional building practices	[25, 51]

Table 1.
Factors hindering the use of green architecture.

It takes a long period to deliver because most of these green products and materials are usually imported. Similarly, Davies and Davies submitted that the unavailability of green building materials and products locally in countries such as Nigeria serves as a major barrier to the adoption of sustainable architecture [51]. As such, built environment professionals find it impossible to relinquish the conventional methods of building construction. Therefore, indigenous companies that manufacture green building products and materials are needed for the growth and development of green architecture.

The various factors hindering the use of green architecture identified from the numerous literature reviewed above are summarised in **Table 1**.

3. Research methodology

The research primarily employed the deductive method to achieve the research aim. Data obtained from secondary sources such as textbooks, journals, workshops/seminars/conference papers, magazines, newspapers and internet sources, etc. were used to review works of literature on sustainable architecture and develop a structured questionnaire that identifies the various factors hindering its use. The questionnaire was piloted on four respondents who are the Architects, Builders, Engineers, and other construction professionals to establish the various factors inhibiting the use of green architecture in Nigeria. Comments and observations from the preliminary survey were incorporated into the final questionnaire. The study employed a structured questionnaire administered to various built environment professionals in Nigeria. The structured questionnaire contained two parts. The first part presented the respondents' profile, made up of educational background, years of experience, and experience level with building construction. In contrast, the second part presented eight factors hindering the use of green architecture in Nigeria, which were deduced from literature. The various responses on each factor were placed on a five-point Likert scale. At the same time, the respondents were asked to indicate their degree of agreement with the factors on the Likert scale in which *Five (5) represents strongly agreed, four (4) represents agreed, three (3) represents undecided, two (2) represents disagreed, one (1) represents strongly disagreed,* and their values were ranked in order of importance to outline the level of significance of each factor. The frequency, percentage count, mean item score, and the relative importance index (RII) were used for data analyses and to indicate the relative importance of each variable, contributing to the factors. The relative importance index (RII) was computed as established by [59] in Eq. (1) below:

$$RII = \frac{5n5 + 4n4 + 3n3 + 2n2 + n1}{5(n5 + n4 + n3 + n2 + n1)} \quad (1)$$

where
n1—number of respondents who responded with 'strongly disagree'.
n2—number of respondents who responded with 'disagree'.
n3—number of respondents who responded with 'undecided'.
n4—number of respondents who responded with 'agree'.
n5—number of respondents who responded with 'strongly agree'.

4. Results and discussion

4.1 Respondents' profile

The Primary data for this research work was obtained through online and manually distributed questionnaires to build environment professionals in Nigeria's construction industry, and the responses gathered were presented in **Table 2**. A total of 200 questionnaires were randomly distributed to various professionals in Nigeria's construction industry, out of which only 112 responded. 38.39% of the respondents are architects, 18.75% are builders, 32.14% are engineers, while 10.71% had other educational disciplines in the construction industry. The results revealed the average years of experience of the respondents to be between 5 and 15 years, implying that all the respondents have significant years of experience in the construction industry.

4.2 Perception of respondents on factors inhibiting the use of green architecture in Nigeria

Table 3 presents the assessment of the various factors hindering the use of green architecture in Nigeria based on a 5-point Likert scale.

The factors hindering the use of green architecture, as identified from the literature and corroborated by various built environment professionals, were ranked according to their Mean Index Score (MIS) and Relative Importance Index (RII) as indicated in **Table 3** and **Figure 1**, respectively. The findings of the empirical analysis revealed that the most significant factors that hinder the adoption and use of green architecture in Nigeria are strongly linked to cost, whereas the least significant factors are closely linked to knowledge and technical capacity. Results from the study further revealed that about 70% of the respondents agreed that lack of financial and non-financial incentives that promote green architecture hinders its adoption and use in Nigeria, whereas 64% of the respondents agreed that insufficient support from the government hinders the adoption of sustainable architecture in Nigeria. The result is consistent with the assertion made by Ndihokubwayo et al. and Chan et al., who believed that most construction industries in developing countries are lagging in terms of providing support and incentives that encourage the incorporation of green architecture concepts into practice [39, 40]. As such, it is now evident that with the outcome of this study, Nigeria is not an exception. Therefore, the active participation of the government in the provision of incentives will undoubtedly encourage the promotion and adoption of green architecture for use in Nigeria. The finding of this study further supports the claim of Ametepey et al. and Du Plessis,

	Frequency	Percentage (%)
Sex		
Male	89	79.46
Female	23	20.54
Total	112	100
Profession		
Architect	43	38.39
Builder	21	18.75
Engineer	36	32.14
Others	12	10.71
Total	112	100
Education history		
Polytechnic (HND)	18	16.07
University (B.Sc.)	62	55.36
University (M.Sc.)	32	28.57
Total	112	100
Years of experience		
Less than 5 years	33	29.46
5–10 years	46	41.07
10–15 years	17	15.18
15–20 years	10	8.93
More than 20 years	6	5.36
Total	112	100

Table 2.
Respondents' profile data.

who claimed that the majority of built environment professionals are not interested in green technological changes [27, 32]. This results consequently in cultural and social opposition due to the general lack of demand from clients and other construction industry stakeholders. Therefore, the progress of green/sustainable architecture in Nigeria depends heavily on the willingness of built environment professionals and other construction industry stakeholders to be fully committed to green technological change and to work towards the acceptance of its principles in Nigeria. Other factors that were found to hinder the use of green architecture in Nigeria are high perceived costs associated with green architecture and preference for traditional building practices, with 57 and 56% of respondents agreeing with such factors. This result correlates with the submission of Hakkinen and Belloni and Ametepey et al., who believe that the cost of sustainable architecture is high with higher investment costs than conventional construction [26, 54]. As such, it results in conventional construction methods being preferred. The three least significant factors hindering the use of green architecture as perceived by the respondents include Technology/Capacity barrier among built environment professionals, Limited knowledge and awareness regarding the economic benefits and prospects of green architecture and the Unavailability of local Green building materials having a Mean Index Score (MIS) of 3.27, 2.93 and 2.56, respectively. This suggests that most respondents believe that Nigeria has the capacity and knowledge of green architecture, as well as

S/N	Factors	Responses					Total (N)	ΣW	Mean (ΣW/N)	RII	Rank
		Strongly Agreed	Agreed	Undecided	Disagreed	Strongly Disagreed					
1.	Technology/capacity barrier among built environment professionals	26	35	12	21	18	112	366	3.27	0.65	6th
2.	Cultural and social resistance by various stakeholders	28	41	9	19	15	112	384	3.43	0.69	3rd
3.	High perceived cost associated with green architecture	25	39	14	20	14	112	377	3.37	0.67	4th
4.	Lack of financial and non-financial incentives that promote green architecture	36	42	9	15	10	112	415	3.71	0.74	1st
5.	Limited knowledge and awareness regarding the economic benefits and prospects of green architecture	20	27	11	33	21	112	328	2.93	0.59	7th
6.	Unavailability of local Green building materials	13	18	15	39	27	112	287	2.56	0.51	8th
7.	Insufficient support from the government	38	34	7	19	14	112	399	3.56	0.71	2nd
8.	Preference for conventional building practices	27	36	10	22	17	112	370	3.30	0.66	5th

Table 3.
Perception of respondents on factors hindering the use of green architecture in Nigeria.

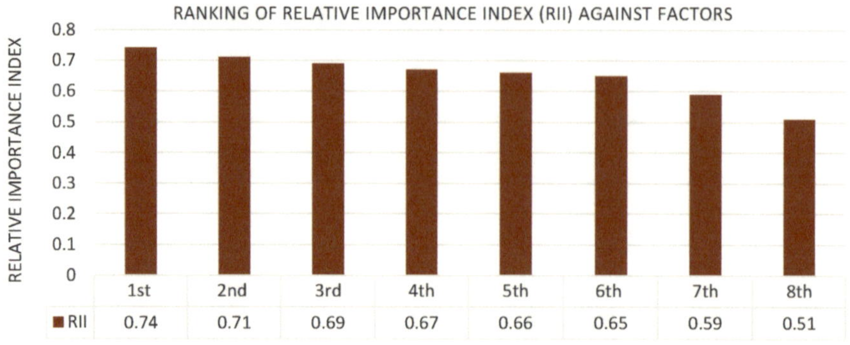

Figure 1.
Ranking of factors hindering the use of green architecture. 1st, lack of financial and non-financial incentives that promote green architecture; 2nd, insufficient support from the government; 3rd, cultural and social resistance by various stakeholders; 4th, high perceived cost associated with green architecture, 5th, preference for conventional building practices; 6th, technology/capacity barrier among built environment professionals; 7th, limited knowledge and awareness regarding the economic benefits and prospects of green architecture; 8th, Unavailability of local Green building materials.

available green building materials locally. However, other factors identified above inhibit its adoption and usage.

5. Conclusion

The concept of green/sustainable architecture is relatively new in the building construction industry and has been asserted as a way of achieving environmental sustainability. However, the adoption and utilisation of its various principles in a developing country such as Nigeria are still quite low, with various factors identified to be hindering its usage. Therefore, these factors need to be addressed in order to promote the adoption and utilisation of green architecture. For that reason, this study examined the various factors hindering the use of green architecture in Nigeria. In order to achieve this aim, a comprehensive empirical review of related literature was conducted, and questionnaires administered to various built environment professionals. Findings from this study revealed that the three most significant factors hindering the use of sustainable architecture in Nigeria are lack of financial and non-financial incentives that promote green architecture, insufficient support from the government as well as cultural and social resistance by various stakeholders. This study, therefore, recommends that since lack of financial and non-financial incentives that promote green architecture and insufficient support from the government, which are government-related are the most significant factors hindering the adoption and use of the various principles of green architecture in Nigeria. Therefore, the active involvement of the government is necessary for overcoming these challenges. This can be done through the introduction of laws, the provision of incentives, and the development of a framework that encourages the implementation of sustainable architecture practices in the building construction industry. This study was not only able to contribute and fill existing knowledge about green/sustainable architecture in Nigeria, but also provides the most significant factors that hinder its acceptance and use. The findings of this study will substantially help to mitigate the challenges of green/sustainable architecture adoption in Nigeria and subsequently contribute to environmental sustainability.

Author details

Auwalu Faisal Koko[1]* and Muhammed Bello[2]

1 College of Civil Engineering and Architecture, Zhejiang University, Hangzhou, China

2 Kaduna Polytechnic, Kaduna, Nigeria

*Address all correspondence to: faisal.koko@yahoo.com

References

[1] Paola S. Strategies for Sustainable Architecture. New York: Taylor & Francis; 2006

[2] Roy M. Importance of Green Architecture Today. Kolkata, India: Department of Architecture, Jadavpur University; 2008

[3] Lehrer D. Facility Design and Management Handbook. New-York: McGraw-Hill; 2001

[4] Thomas R. The Principles of Green Building Design; INTD 62 Spring 2009: Architect and Niklas Spitz Monterey Peninsula College; 2009

[5] Brenda V, Robert V. "Principles of Green Architecture" from Green Architecture (1991). In: Stephen M, Beatley T, editors. The Sustainable Urban Development Reader. London/ New York: Taylor and Francis; 2014

[6] Otegbulu AC. Economics of green design and environmental sustainability. Journal of Sustainable Development. 2011;**4**:240-248. DOI: 10.5539/jsd.v4n2p240

[7] William EK. Engineering for sustainable development and the common good. Current Issues in Catholic Higher Education. 2006;**25**:43-64

[8] Jeremy DR. Sustainability and civil engineering. In: Ohio Valley Student Conference Paper Presentation Competition. Kentucky, USA: Western Kentucky University; 2009

[9] Gottfried DA. Sustainable Building Technical Manual: Green Building Design, Construction, and Operation. 2019. Available from: https://www.buildinggreen.com/newsbrief/sustainable-building-technical-manual-green-building-design-construction-and-operation [Accessed: 20 November 2019]

[10] Adesoji BS. Despite Setbacks, Nigeria's Real Estate Investors to Expect Windfall in 2019. Available from: https://nairametrics.com/2019/03/13/despite-setbacks-nigerias-real-estate-investors-to-expect-windfall-in-2019/ [Accessed: 06 April 2020]

[11] Gidado Dalibi S, Feng J, Shuangqin L, Sadiq A, Bello B, Danja I. Hindrances to Green Building Developments in Nigeria's Built Environment: "The Project Professionals' Perspectives." In Proceedings of IOP Conference Series: Earth and Environmental Science, Suzhou, China, 05/01

[12] U.S. Environmental Protection Agency. Green Building Basic Information. Available from: http://www.epa.gov/greenbuilding/pubs/about.htm [Accessed: 12 December 2019]

[13] Elshimy H. Green building as concept of sustainability sustainable strategy to design office building. Applied Sciences and Engineering. 2015;**2**:41-54

[14] Kadiri KO. Planning sustainable and livable cities in Nigeria. Research Journal of Social Sciences. 2006;**1**:40-50

[15] Salama AM, Ashukaikhat M. A trans-disciplinary approach for a comprehensive understanding of sustainable affordable housing. Global Built Environment Review. 2006;**5**:35-50

[16] Goldemberg J, Johansson TB, Reddy AKN, Williams RH. Energy for a Sustainable World. New York: Wiley; 1987

[17] Unalan H, Tokman LY. Building sustainable architectural design: A renovation project. Anadolu University

Journal of Science and Technology A: Applied Sciences and Engineering. 2011;**12**:129-157

[18] Dincer I, Rosen MA. Exergy: Energy, Environment and Sustainable Development. 2nd ed. Amsterdam: Elsevier Science Publishers; 2007

[19] Sirinja M. Necessity of Sustainability in architectural practices for achieving sustainable development. International Journal of Science and Technology. 2013;**2**:583-587

[20] Khalfan M, Noor MA, Maqsood T, Alshanbri N, Sagoo A. Perception towards sustainable construction amongst construction contractors in State of Victoria, Australia. Journal of Economics Business and Management. 2015;**3**:940-947. DOI: 10.7763/JOEBM.2015.V3.313

[21] Choi C. Removing market barriers to green development: Principles and action projects to promote widespread adoption of green development practices. Journal of Sustainable Real Estate. 2009;**1**:107-138. DOI: 10.5555/jsre.1.1.7736t75545j02597

[22] Issa MH, Rankin JH, Christian AJ. Canadian practitioners' perception of research work investigating the cost premiums, long-term costs, and health and productivity benefits of green buildings. Building and Environment. 2010;**45**:1698-1711. DOI: 10.1016/j.buildenv.2010.01.020

[23] Hamidi B. A Green Cost Allocation Model for Office and Commercial Buildings in Malaysia. Johor Bahru, Malaysia: Universiti Teknologi Malaysia; 2010

[24] Means RS. Green Building Project Planning and Cost Estimating. London, U.K: Reed Construction Data Publisher; 2002

[25] Robichaud LB, Anantatmula VS. Greening project management practices for sustainable construction. Journal of Management in Engineering. 2010;**27**:48-57. DOI: 10.1061/(ASCE)ME.1943-5479.0000030

[26] Hakkinen T, Belloni K. Barriers and drivers for sustainable building. Building Research and Information. 2011;**39**:239-255. DOI: 10.1080/09613218.2011.561948

[27] Du Plessis C. Agenda 21 for Sustainable Construction in Developing Countries: a Discussion Document. Pretoria: CSIR Building and Construction Technology; 2002

[28] Opoku A, Ahmed V. Leadership and Sustainability in the Built Environment. London: Taylor and Francis; 2015

[29] Rydin Y, Amjad U, Moore S, Nye M, Withaker M. Sustainable Construction and Planning. The Academic Report. In: The LSE SusCon Project, Centre for Environmental Policy and Governance (CEPG). Edn. London: London School of Economics; 2006

[30] International Labour Office. Greening of the building sector is held back by skill shortages. In: Research Brief for Skills and Occupational Needs in Green Building. Geneva: European Union; 2011. pp. 1-12

[31] Ahn YH, Pearce AR, Wang Y, Wang G. Drivers and barriers of sustainable design and construction: The perception of green building experience. International Journal of Sustainable Building Technology and Urban Development. 2013;**4**:35-45. DOI: 10.1080/2093761X.2012.759887

[32] Ametepey O, Aigbavboa C, Ansah K. Barriers to successful implementation of sustainable construction in the Ghanaian Construction Industry. Procedia Manufacturing. 2015;**3**:1682-1689. DOI: 10.1016/j.promfg.2015.07.988

[33] Kibert CJ. Sustainable Construction: Green Building Design and Delivery. New Jersey, USA: John Wiley & Sons; 2008

[34] Byrd H, Leardini P. Green buildings: Issues for New Zealand. Procedia Engineering. 2011;**21**:481-488. DOI: 10.1016/j.proeng.2011.11.2041

[35] Hydes KR, Creech L. Reducing mechanical equipment cost: The economics of green design. Building Research and Information. 2000;**28**:403-407. DOI: 10.1080/096132100418555

[36] Shi Q, Zuo J, Huang R, Huang J, Pullen S. Identifying the critical factors for green construction—An empirical study in China. Habitat International. 2013;**40**:1-8. DOI: 10.1016/j.habitatint.2013.01.003

[37] Azeem S, Naeem MA, Waheed A, Thaheem MJ. Examining barriers and measures to promote the adoption of green building practices in Pakistan. Smart and Sustainable Built Environment. 2017;**6**:86-100. DOI: 10.1108/SASBE-06-2017-0023

[38] Oguntona O, Akinradewo O, Ramorwalo D, Aigbavboa C, Thwala W. Benefits and drivers of implementing green building projects in South Africa. Journal of Physics Conference Series. 2019;**1378**:032038. DOI: 10.1088/1742-6596/1378/3/032038

[39] Ndihokubwayo R, Crafford G, Buys F. Consultant team members' performance evaluation against incentives towards the achievement of green building principles. In: Proceedings of Green Vision 2020, SACQSP Research Conference; Cape Town, South Africa; 20-21 June, 2013

[40] Chan APC, Darko A, Olanipekun AO, Ameyaw EE. Critical barriers to green building technologies adoption in developing countries: The case of Ghana. Journal of Cleaner Production. 2018;**172**:1067-1079. DOI: 10.1016/j.jclepro.2017.10.235

[41] Darko A, Chan A. Review of barriers to green building adoption. Sustainable Development. 2016;**25**:167-179. DOI: 10.1002/sd.1651

[42] Darko A, Chan APC, Owusu-Manu D-G, Ameyaw EE. Drivers for implementing green building technologies: An international survey of experts. Journal of Cleaner Production. 2017;**145**:386-394. DOI: 10.1016/j.jclepro.2017.01.043

[43] Arditi D, Yasamis F. Incentive/disincentive contracts: Perceptions of owners and contractors. Journal of Construction Engineering and Management. 1998;**124**:361-373

[44] Serpell A, Kort J, Vera S. Awareness, actions, drivers, and barriers of sustainable construction in Chile. Technological and Economic Development of Economy. 2013;**19**:272-288. DOI: 10.3846/20294913.2013.798597

[45] Sodagar B, Fieldson R. Towards a sustainable construction practice. Construction Information Quarterly. 2008;**10**

[46] Bernstein HM. World Green Building Trends: Business Benefits Driving New And Retrofit Market Opportunities. In: Over 60 Countries. New York: McGraw-Hill Construction; 2013

[47] Abidin NZ, Yusof N, Awang H. A foresight into green housing industry in malaysia. In: World Academy of Science, Engineering And Technology. International Journal of Environmental, Chemical, Ecological, Geological and Geophysical Engineering. 2012;**6**:373-381

[48] Nguyen HD, Nguyen LD, Chih YY, Le-Hoai L. Influence of participants'

characteristics on sustainable building practices in emerging economies: Empirical case study. Journal of Construction Engineering and Management. 2017;**143**:05017014

[49] William K, Dair C. What is stopping sustainable building in England? Barriers experienced by stakeholders in delivering sustainable development. Sustainable Development. 2007;**15**:135-147

[50] Alabi AA. Comparative study of environmental sustainability in building construction in Nigeria and Malaysia. Journal of Emerging Trends in Economics and Management Sciences. 2012;**3**:951-961

[51] Davies OOA, Davies IOE. Barriers to implementation of sustainable construction techniques. MAYFEB Journal of Environmental Science. 2017;**2**:1-9

[52] Aktas B, Ozorhon B. Green building certification process of existing buildings in developing countries: cases from Turkey. Journal of Management in Engineering. 2015;**31**:05015002. DOI: 10.1061/(ASCE)ME.1943-5479.0000358

[53] Shen L, Zhang Z, Long Z. Significant barriers to green procurement in real estate development. Resources, Conservation and Recycling. 2017;**116**:160-168. DOI: 10.1016/j.resconrec.2016.10.004

[54] Ametepey SO, Gyadu-Asiedu W, Assah-Kissiedu M. Sustainable construction implementation in Ghana: Focusing on awareness and challenges. Civil and Environmental Research. 2015;7:109-119

[55] Hankinson M, Breytenbach A. Barriers that impact on the implementation of sustainable design. In: Proceedings of Cumulus Conference; Helsinki, Finland; 24-26 May, 2012. pp. 1-11

[56] Wadu Mesthrige J, Kwong Ho Y. Criteria and barriers for the application of green building features in Hong Kong. Smart and Sustainable Built Environment. 2018;7:251-276. DOI: 10.1108/SASBE-02-2018-0004

[57] Wang W, Zhang S, Su Y, Deng X. Key factors to green building technologies Adoption in developing countries: The perspective of Chinese designers. Sustainability. 2018;**10**:4135. DOI: 10.3390/su10114135

[58] Darko A, Zhang C, Chan APC. Drivers for green building: A review of empirical studies. Habitat International. 2017;**60**:34-49. DOI: 10.1016/j.habitatint.2016.12.007

[59] Aziz N, Zain Z, Mafuzi R, Mustapa A, Najib N, Lah N. Relative importance index (RII) in ranking of procrastination factors among university students. In: Proceedings of International Conference on Applied Science and Technology (ICAST'16); 2016

Chapter 3

Road-Mapping for a Zero-Carbon Building Stock in Developed and Developing Countries

Dirk Schwede

Abstract

Given the global climate crises, the enormous construction activity and the rising demand for comfortable living spaces around the world, it is not only the task for today to explore the feasibility of zero-energy buildings based on advanced technology concepts, but also the task for a zero-carbon future to transform the entire building stock. This chapter explores an integrated road-mapping approach to guide the various relevant levels of global, regional and national governance, on sector level as well as on the level of individual buildings. It will explore how key technologies, individual building configurations, infrastructure and the governance framework can be strategically developed in specific market contexts to achieve ambitious performance goals in the given time frame. It also introduces the concept of individual building renovation roadmaps and design features to be prepared in new and existing buildings to enable the retrofit of key technologies when they become economically and technically feasible in the given market. The roadmap approach with a clear performance target and a mid- and long-term vision is paramount since market conditions do not exist yet to implement such buildings in all market situations today. The text presents the concept of transformation roadmaps on the various levels of implementation and introduces examples.

Keywords: transformation, roadmap, zero-carbon building stock, governance framework, developing markets, life-cycle approach, future-proof design

1. Introduction

We do know today how to design and how to build near-zero-emission buildings. Even plus-energy building performance can be achieved [1] by the integration of highly energy-efficient building design, advanced energy systems and solar energy applications. Under most climate conditions in the world, the annual energy demand and energy generation in buildings can be balanced in the buildings' annual performance, if advanced design principles and technologies are employed, and the ratio between the usable floor area and the solar roof area is not too large [2]. Technically, such performance can be achieved in newly designed, but also in retrofitted buildings. The key technologies are available and technical development is progressing to make low energy performance buildings more and more feasible in future in all climate zones worldwide.

However, although the global climate crisis is apparent, today only a tiny percentage of buildings is built as near-zero-emission buildings and the existing building stock is not performing anywhere near the required efficiency. In the European Union, at least 75% of the buildings that exist today will still exist in 2050, and only 20–25% will be built new in this period [3]. Therefore, besides the design of new buildings, the retrofit of existing buildings towards zero-emission performance and plus-energy performance is to be pursued throughout the entire building stock worldwide.

While the 2050-building stock is already existing today in the developed countries to a large extent, in other places with currently ongoing urbanisation, new buildings need to be constructed that are able to fulfil the future requirements for low-carbon performance. The low-emission goal is the same, but the starting points and boundary conditions in the various countries are significantly different [4]. Often in rapidly developing economies, where most of the construction activity takes place today, other priorities than the objective to build low-carbon buildings drive this development. Also, the governance framework and the building ordinance are not developed yet, and the local market capacities are not as advanced as necessary to build zero-carbon buildings at the current stage. For example, in many countries, the necessary testing facilities to ensure compliance with high-performance specifications for material, building systems and technical appliances do not exist today [5]. The broad implementation of zero-carbon buildings is also hindered by the available budget for energy-saving investment and other spending priorities. Nevertheless, many clients are aware today that energy prices will rise in the foreseeable future. Also, the connection between fossil fuel consumption for energy services in the building sector and climate change is well known to the clients and the other stakeholders today so that efforts towards advanced performance targets are not hindered by ignorance anymore, but clearly by other barriers.

While a single pilot project is still useful today, the challenge today lies in the implementation of the low-emission performance concepts in the building stock in the markets worldwide. However, there are surely crucial technologies, which continue to be required to be developed, designed and demonstrated in zero-energy building pilot projects in the various contexts around the world. Meanwhile, the effort to transform the global building stock is becoming more process- and policy-oriented, rather than being targeted towards the application of specific technologies and the integration of technology systems on building level. Today, the available technologies need to be rolled out for broad application.

2. Contextual conditions

Technically, the fundamental principles for low energy and zero-energy performance of buildings differ in the various climates. The impact of the various components of the building energy balance will result in different configurations of the physical building systems and the building energy systems depending on the local conditions.

It is also evident that some of these technical configurations while being technically feasible based on international best practice are not economically feasible today and might also overstress the local market capabilities, the skills of the workforce and economic capabilities and the willingness of the local clients to invest in low-carbon design. However, as the building sector is a significant part of all local economies today, with about 30–40% of the national energy demand in most countries, in order to advance the zero-carbon development, especially the building sector must contribute. The building sectors' development in such contexts needs to be guided since there currently is, and there will be, significant pressure from the urbanisation and the parallel socio-economic development in the available time frame. In order to avoid login effects through unsuitable building design and the omission of required technical provisions under these conditions, the design must be prepared with appropriate foresight to achieve the intended performance if not now then at least in the long run before 2050.

3. Development of contextual conditions

In the assessment of advanced energy concepts besides these conditions of the application of key technologies in each market, also other factors of the market context need to be observed. For instance, under the usual accounting methodology to offset unavoidable emissions, such as for heating in winter or for electricity use in the night with credits for energy generation in times of availability of renewable sources, the assessment will change depending on the change of the emission factors of the delivered energy to the site. Therefore, in a context where clean electricity is supplied by the local grid, building integrated renewable energy systems can contribute less to reduce the carbon account of a building than in a context with emission-intensive energy infrastructure.

As we can assume that the energy mix will move towards a higher percentage of renewable energy sources through the transformation of the national energy infrastructure in many places, the building-related balance calculation needs to be updated regularly over time towards the intended 2050 performance. In case conventional energy sources are used for heating or domestic hot water generation, the results will show that the building's low-carbon performance decreases if the energy consumption is not decreased at the same time, or the building-integrated renewable energy supply is not increased. In other words in situations with emission factors of 600 g/kWh in the local energy mix, 1 kWh renewable energy produced inside the building will off-set double the amount of CO_2 emission from fossil energy carriers than in situations with a carbon-emission factor of 300 g/kWh.

In consequence, while it is essential to reduce the carbon emission factor of the local electricity grid as a strategy on the national and local municipality level, the zero-carbon strategy on building level must reduce the use of fossil energy carriers for building operation through energy-saving measures and in the last consequence by replacing these fossil applications with clean alternatives in the course of the buildings' life cycle.

The feasibility of such alternatives currently rests with the ability to align the time profiles of energy demand and energy supply from alternative energy sources. As renewable sources depend on environmental processes such as wind and solar radiation, storage technologies are needed. There are three possible solutions to this problem, which differ in their demand for infrastructure and investment:

1. Installation of energy storage solutions as public infrastructure to buffer clean feed-in energy centrally and then to supply seamlessly back to the end-users.

2. Installation of energy storage applications as part of the building system to buffer available clean energy until it is used inside the building by the end-user.

3. To adjust the users' demand to the available resources and to accept that not all demands can be met in the building at all times.

These three strategies can be combined in the building's renovation roadmap. Design in a developing market might first rely on a sufficiency approach (option 3), in which the building design can support the best possible functionality and reduce discomfort, but not guarantee the complete set of performance, which could be achieved when energy would be available at all times. Then, option 1 or 2 could be installed later in the buildings' life cycle as a more functional solution when funds and other capacities are available. Also, in the case of the more extended future where central solutions are developed (option 1), an individual approach (option 2) might only be used as an intermediate solution. In all these cases, the original building design needs to be prepared from the beginning to accommodate the necessary change.

A similar situation can be observed in many developing countries for the installation of room conditioning systems for cooling and heating. Traditionally, the building users are used to free-running conditions. In such situations, the users have to adapt to the climatic conditions in order to find comfort. At a later stage in the buildings' life cycle, the comfort demand changes and technology becomes affordable in the given market. In consequence, the retrofit of conditioning systems becomes necessary. As we can foresee such change of demands, the retrofit should be factored into the design from the beginning. The initial design should include sufficient installation spaces in appropriate locations for the indoor and outdoor units, and airtightness of the building envelope and appropriate zoning of the interior spaces must be prepared.

4. Future-proof design

A future-proof design of buildings takes the long-term goals and developments into account at the time of designing. It is not only considered what can be implemented today, but what is to be added and retrofitted in future to complete the building system in order to achieve the intended performance. Future-proof design can consider future energy retrofit measures towards the projected zero-emission performance, but also other objectives, such as barrier-free design or the extension of the floor space area in case that the occupants' living situation is changing in the building's lifespan.

For instance, "solarisation" of master plans is an objective of master plan development to allow for the installation of functional and economically feasible solar systems. This solarisation would include the orientation of unshaded roof areas to the south, the roof slope, the size and dimensions of roof areas so that solar systems can be fit as well as the location of the buildings in the neighbourhood in case these roofs are planned to be part of a district heating system in future [6]. Such foresight in master planning is essential for solar thermal systems and also to achieve the highest electricity harvest in case of PV installations. Hence, in order to prepare our buildings for future retrofit of solar systems, provisions need to be made. The requirements thereby depend on the context. In situations with low module prices and high energy tariffs (as in Germany), the design of PV systems is not as restricted anymore by economic constraints and therefore building design for integrated PV systems is freer than in situations with high technology prices compared to the available budget and low saving potentials due to low energy tariffs in a given market, such as in Vietnam.

Future-proof design will make provisions for installation of available technologies that are not economically feasible at the time of the first construct, due to high installation cost with the available budget or uneconomic performance due to low energy prices. In many countries and especially in the emerging economies,

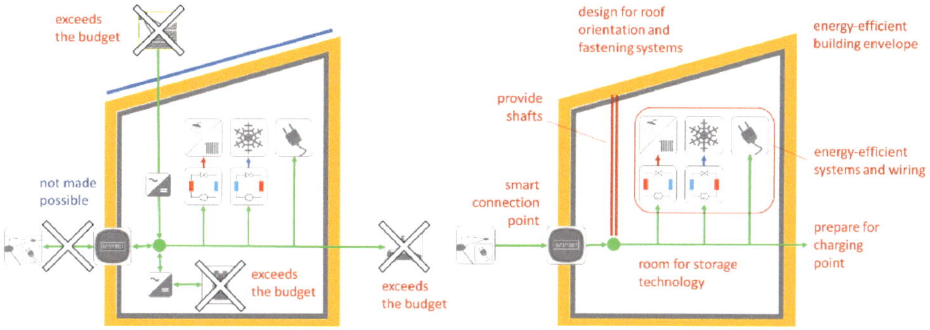

Figure 1.
Zero-emission-ready building design.

energy tariffs are subsided for economic and social reasons. It can be expected that subsidies are reduced and that energy rates will increase and that at a point in time in the building's lifespan retrofit of energy-saving measures and renewable energy applications will become more feasible and convincing than they are today. An example of future-proof design is given in **Figure 1**.

Future-proof design can also be made in the expectation of future technologies, such as new energy storage systems or new low-carbon fuels. As such systems move up the technology readiness levels in research and development, they will become available for installation in future.

5. Road-mapping

5.1 Introduction

A technology roadmap is an instrument to outline the expected future development and the boundary conditions of such development. The application of roadmaps built on the hypothesis that the future does not simply happen but can be constructed with a view towards a desirable future [7]. A roadmap helps to align short-term targets with long-term goals and directions [8]. The roadmap also helps to understand the context of technical developments better. A technology roadmap will document the current context and will draft the desired future performance. It will then develop a pathway from the current situation towards the intended future performance. In the original form, a roadmap requires a graphical representation and the time axis to depict the required steps in their sequence towards the desired state. Today, besides the graphical representation, table structures and narratives are used in the development and communication of roadmaps. Such roadmap is not static-once developed and then applied until the end of its time frame- but a roadmap design requires mechanisms for its review and continuous improvement [7].

As instruments for planning for the future and for directing the development towards the desired state, roadmaps have been used on various levels in management, economy and policy contexts. Policy roadmaps and sector roadmaps are used widely today to understand the contexts and requirements of transformation processes on the larger scale of countries or worldwide. In companies, roadmaps are used for product planning, the development of capabilities and strategic knowledge assets and to align activities between departments towards a coordinated goal. Phal et al. [7] list various types of roadmaps as given in **Table 1** and their specific purposes, and obviously, roadmaps can be used for further applications.

Purpose	Description
Product planning	A common type of roadmap, aligning technology and product strategy, typically including more than one generation of product
Capability planning	More suited to service-based enterprises, focusing on the insertion of technology into organisational capabilities
Strategic planning	Includes a strategic dimension, supporting the evaluation of different opportunities or threads typically at the business level
Long-range planning (foresight)	Typically developed at the sector or national level incorporating longer time horizons (e.g., industry, science and policy, foresight roadmaps)
Knowledge and asset planning	Focus on aligning knowledge assets and knowledge management initiatives with business objectives
Programme planning	Focuses on the implementation of strategy, to support the management of integrated R&D programmes. This type is more closely related to project planning methods (Gantt charts)
Process planning	Supports the management of knowledge, focusing on the knowledge flows necessary to support a particular process area, such as new product development
Integration planning	Focuses on the integration and/or evolution of technology, and how different technologies combine to form new technologies

Table 1.
Types and purposes of roadmaps, cited from Phal et al. [7, p. 16].

A multilevel approach is required involving the identification of technologies and the technological context, the assessment of its compatibility and the complementary of various technologies and the integration of the technologies into the system as well as the implementation of first implementation instances to introduce new technologies and concepts into a market [9].

Often technology roadmap designs are structured with the layers such as the science layer, the technology layer, the application layer and the market layer (S-T-A-M framework) [7], each addressing one aspect of a product's life cycle. On a higher level of technology application in the design of a governance framework, the layers social context, technology context, economic context, environmental context, political context, legal context and infrastructural factors could be used. A roadmap can be designed to fit the needs of the specific application. In this chapter, a retrofit plan for a low-emission building is to be promoted with layers along with the main components of the energy balance and the elements of the energy concept.

5.2 Applications of roadmaps

Rockström et al. [8] suggest a global roadmap for rapid decarbonisation. The authors remark the model-based decarbonisation strategies often fall short of capturing transformative change and the dynamics of development involving disruption, innovation and nonlinear change in human behaviour. Therefore, they suggest drafting a roadmap towards the achievement of global decarbonisation based on simple principles, such as halving the anthropogenic carbon dioxide emissions every decade. Such a goal will serve as guide rail with increasing ambition every decade to achieve the required change in steps until the long-term goal is reached.

However, the simpler the roadmap to guide policies, the lesser its use to achieve the goal in the field. We need to translate the goals on the global policy level to national goals in every country and sector goals in every sector in every country. Ultimately, these goals must be applied in every facility and case of the building sector for every existing and newly built building. Rockström et al. [8] structure their roadmap framework and suggest developing narratives in the dimensions innovation, institution, infrastructure and investment to understand the required conditions and steps in the various sectors. These narratives could be reviewed and newly aligned to the actual development in regular steps. The narratives developed in the various sectors by the sector stakeholders and experts can be used to align actors and organisations to achieve a common goal through trans-sectoral transformation.

In consequence, not only one broad roadmap is required, but countless individual roadmaps are required towards delivering the individual share to the solution. Since all these individual roadmaps are directed towards the same common goal, actions are aligned and are in the best case supportive to each other.

Benefits can be realised through an interplay of the roadmaps on the various levels. For instance, in many situations, the first movers are essential to start the required market transformation, but these first movers often face disadvantages, such as pioneering costs, reluctance in the market and the investment in early technology generations. In such situations, a roadmap on a higher level can provide security for first movers' investment and activity [1], so that the early majority can follow, and the application is taken up by the mainstream. The long-term framework of the German feed-in tariff is an example of such support. Since the feed-in tariff was guaranteed for 20 years, investors were willing to invest in an early generation of solar photovoltaic systems, which in turn provided the market for first movers in the solar industry in Germany and later worldwide.

Figure 2 depicts the scheme of a layered integrated implementation roadmap towards zero-carbon building performance in 2050. Only the aligned interplay between the layers will allow the stakeholders on the building layer to implement near-zero-emission buildings effectively and to align their efforts towards the same goal. Formulation and communication of the specific roadmaps will allow the individual layers' stakeholders to link into the activity of the other layers in order to contribute to transformative change together.

In the context of building's design and the broader field of urban development, a more elaborate narrative of a desirable future defined by a framework of sustainable qualities and performance characteristics can be formulated. Such a narrative

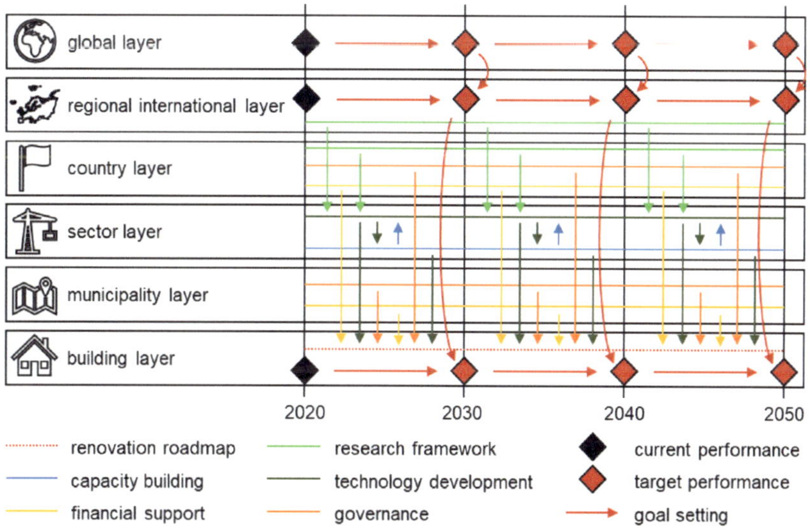

Figure 2.
Schematic depiction of the layers of an integrated implementation roadmap framework towards zero-carbon building performance.

can then give direction for the required development and the communication and alignment of strategies.

In this sense, following the global goal, the requirements set in the Paris Agreement, it is the aim to reduce the carbon emissions of the German building stock towards 0 kg CO_2 emissions till 2050. Such a time perspective provides 30 years to reduce the emission of the building stock by improving the building performance and the performance of the energy infrastructure. However, the activities towards the desired performance must be planned in order to make effective use of the given time. Long-term goal setting will provide security to the market but will also create the required pressure. It will allow all the stakeholders to plan and to start the individual transformation process and will avoid overwhelming the stakeholders by abrupt changes.

It is possible today to evaluate the performance of buildings before they are built, and the performance of different configurations of retrofit measures can be assessed for the development of a retrofit plan, for example through computational simulation studies. To calculate the future performance, the assessor has to make predictions on the development of economic parameters, such as system costs, energy tariffs and operation cost and environmental impacts, such as the composition of the energy mix and the resulting CO_2 emissions.

In case a roadmap approach is applied as part of a governance framework or for performance assessment in a subsidy scheme, these predictions can be defined by the operating body to support the designers and assessors of the individual building projects. Such a definition of the evaluation context is an example, where the country-level roadmap and the building-level roadmap are interlinked. By defining and providing such context information on a higher layer with the mid-term and long-term perspective, the assessment for single building projects but also higher-layer policy interventions can be supported on the lower layers.

In this context, the German sustainable building council has issued a framework for "carbon-neutral buildings and sites" [10]. Herein, carbon accounting rules, rules for CO_2 reporting and carbon management rules (or a climate protection roadmap) are provided. The framework sets current and future individual performance targets of a building or a group of buildings by assessment of the current annual carbon emissions and the target carbon emission of <0 kg CO_2 emission in 2050. The target values in between are retrieved by simple linear interpolation between the current values and the targeted value, as shown in **Figure 3**. If the building performance remains below this line, and the retrofit roadmap is followed, the building can be labelled "carbon-neutral by 2050".

Such a roadmap framework will allow preparing for planned future enhancement of the technical installations in the given time frame until the most ambitious target is reached. Since it sets performance targets in terms of carbon emissions, the framework is not restrictive to any technology and operation strategy.

It will allow identifying technology fields that require or allow the installation of advanced solutions immediately and to schedule the retrofit of other solutions in the building's life cycle. For example, the client could decide to install double-glazed windows today but to schedule the installation of triple-glazed windows in the next renovation cycle in 2040 after the technical lifetime of the first set of windows. At that time, triple-glazed windows might be better available in the local market. Another example could be a building concept that relies on natural ventilation first to reduce the initial cost of construction, but that is prepared for retrofit of a controlled ventilation system with heat recovery in future when such systems are better available from local suppliers. For other technologies with long technical lifetimes, high-performance solutions should be implemented from the beginning. In any case, login effects must be avoided, so that high-performance solutions can

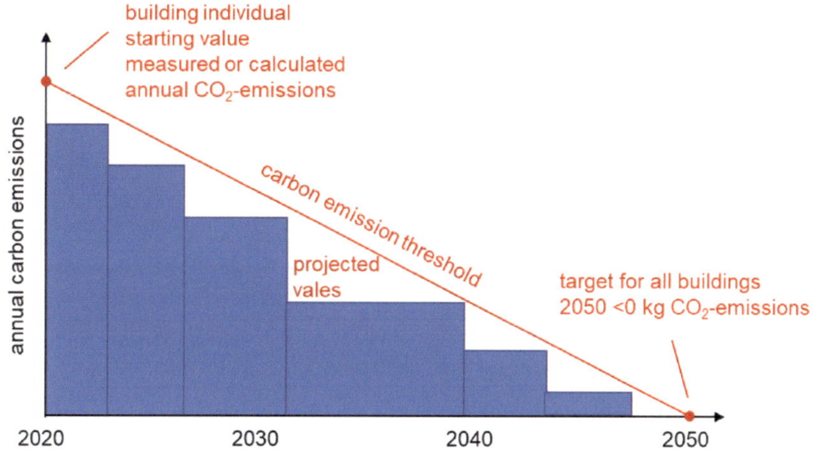

Figure 3.
Roadmap to the climate-neutral building stock. Reproduced from DGNB "Framework for carbon-neutral buildings and sites" [10].

be implemented at a later time in the building's lifetime. Therefore, roof spaces must be designed suitable for solar collectors and shafts must be provided to run the required cables and pipes (see **Figure 1**). For some solutions, road-mapping also gives hints where leapfrogging of inferior technologies is required.

5.3 Modelling for building assessment

Architects and engineers can model the building and its future performance under the local conditions and test and develop such design variants in the computational thermal building simulation model today. In contrast to the usual application of computational simulation models, the optimisation target is first set to <0 kg CO_2 emission with the anticipated boundary conditions of 2050. A building concept with the desired performance and design qualities in 2050 is developed, and the design is detailed, although not all planned systems are to be installed at the time of first construction. The design is then reduced step by step and solutions for the later retrofit towards 2050 performance are developed. The technical systems are evaluated based on their technical lifetime and the required maintenance, and replacement cycles are considered in roadmap development, thereby retrofit, and energy upgrade steps are scheduled in the period till 2050. The performance of each retrofit step is simulated and evaluated with the anticipated boundary conditions at that time.

In order to apply such a roadmap approach, the assessor will need to make predictions of the development of economic parameters, such as installation costs, energy tariffs and operation costs. The assessor will also have to make predictions on future technical development, the technical and economic lifetime of systems, as well as on the development of the market context. Parameters such as labour cost for installation and maintenance work as well as the available skills for technology installation and maintenance need to be considered. Many of these required assumptions can be provided by central bodies, such as the government, administrations, funding institutions or professional associations to support strategic road-mapping on building level towards the international climate protection goals.

Such central generation and provision of required data would reduce the individual degrees of freedom in modelling and thereby make the modelling results more comparable. The boundary conditions can be supplied from higher layer roadmaps and agreed and consolidated research results. For instance, the German Ministry of Construction has recently provided new test reference year datasets for the current climate conditions, but also for the climate in 2050 including the predicted climate change effects to be used in thermal simulation studies for future situations [11].

5.4 Application in the building sector

An example for the application of roadmaps is the "individual renovation roadmap" (individual Sanierungsfahrplan, iSFP) as it is currently introduced by the German Federal Ministry for Economic Affairs and Energy (BMWE) for the application in energy-retrofit projects in Germany [12]. In future, an iSFP will be the basis for a financial support scheme for the retrofit of building projects in Germany. Before isolated retrofit measures are planned, an individual renovation roadmap is drafted to define the intended development of the building's functions and performance. In this way, it is possible to avoid lock-in effects and to make use of synergies in the energy system of the building. The method is explained in detail in the "Handbuch für Energieberater" [12] in relation to the existing building assessment framework in Germany.

Also, the subsidy schemes in Germany support the preparation for renewable energy retrofit. In one programme, investment grants are provided for "renewable ready" condensing boiler systems in case these systems are prepared for later renewable energy integration with prepared additional fittings for necessary piping and functions in the control system. Investment grants are paid to the investor if the renewable energy system is installed within 2 years after the installation of the heating system.

Also, the US Department of Energy has developed a "Zero Energy Ready Home" [13] building labelling system, which builds on top of the existing and forthcoming low energy building requirements and requires the client to prepare for the later retrofit of renewable energy systems. Within the scheme, the client has to demonstrate that the maximum allowable loads of the roof structure are sufficient for the installation of PV solar collectors, that the conduits run from the roof to the dedicated location of the inverter and that the inverter can be connected to the electrical panel and circuit breakers are prepared for installation.

Since in many situations, the current building practice is too remote from the intended performance, such as plus-energy performance or zero-carbon performance, a roadmap approach will be instrumental for the development of the building stock in many countries.

5.5 Steps in the individual renovation roadmap

In summary, the following will draft the steps in the integrated and the individual renovation roadmap. The concept is illustrated in **Figure 4**. It is suggested to develop narratives and technical concepts for the state of the art, the goal in 2050, the situation today and of intermediate steps.

5.5.1 State of the art

In the development of roadmaps for a zero-emission building in a specific context, state of the art needs to be reviewed and currently available advanced options for retrofit are to be listed. Thereby the commonly used technologies, as well as the local and international front runner technologies, need to be addressed to benchmark the current and the possible future performance.

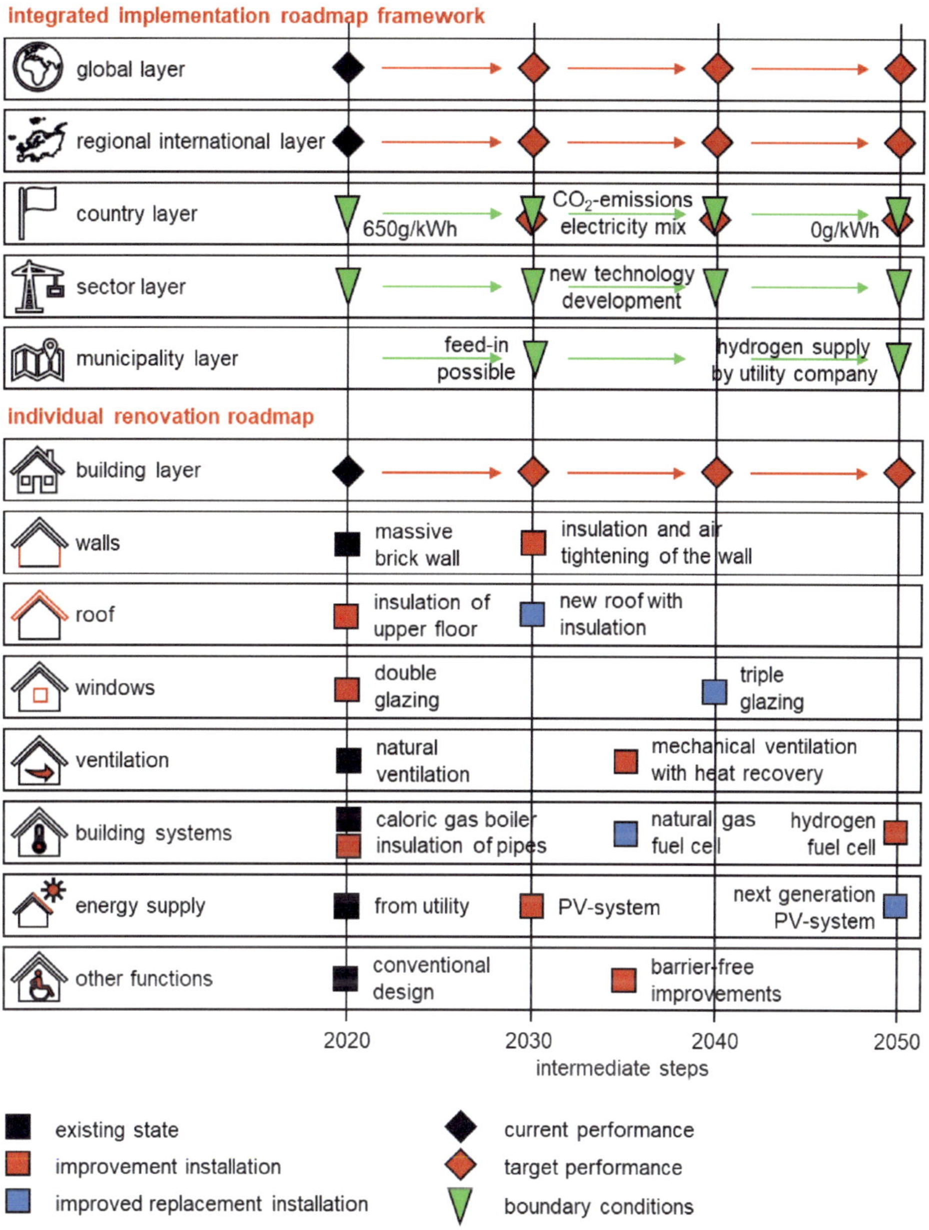

Figure 4.
Illustration of individual renovation roadmap under the influence of the integrated implementation roadmap layers.

The review of state-of-the-art technologies will inspire the designers for the design of the building today and envision the developments of potential key technologies in future. Thereby, a front-runner technology review and the review of technology development, which are currently still on the low end of the technology-readiness level scale of research projects, are very informative.

5.5.2 2050

The performance goal defines the "2050" narrative and the intended performance. In the case of the DGNB roadmap to climate-neutral building stock, it is a total annual carbon-emission of 0 kg. Also, the technical configuration of the building and the boundary conditions for the calculation need to be specified in order to be useful for the building owner. The configuration able to achieve the intended performance can be determined through computational simulation. The technical description is the target for the successive retrofit.

5.5.3 Situation today

Today's state of the building is either in case of building retrofit projects the current condition of a building at the time of assessment or it is the state of a new building at the time of construction in case of a newly built building. This condition is documented as the starting point for roadmap development.

In order to design a "climate-neutral by 2050" building, it is advisable to determine the 2050-configuration first and then to subtract the elements until construction is technically and economically feasible at the current time. The subtraction of elements can be due to financial constraints, or due to technical availability or market maturity or also due to changing demand. Obviously, it is not possible to predict the future precisely, but the road-mapping approach will help to avoid login effects and allow for the design of future-proof buildings.

5.5.4 Intermediate steps

The intermediate steps are defined by the current state and the intended state in 2050. In the method introduced earlier, these intermediate steps are determined by subtraction components from the high-performance configuration. In this process, the technical lifespans of the building components and service systems are considered. For example, double-glazed windows installed in 2020 are exchanged for triple-glazed windows at the end of their lifespan in 2040 to achieve the intended performance in 2050.

6. Conclusion

The chapter has introduced the concept of future-proof design and integrated implementation roadmaps towards the step-by-step achievement of zero-carbon performance in 2050. The application of these concepts has been discussed on the level of governance and the level of single building projects as tools for goal setting and strategic development of future-proof building designs. While on building level, components of the building and energy concepts are addressed in a renovation roadmap; on the higher international, national, municipal and building sector layers, support schemes for research, technology development, capacity building and financial support are introduced in an implementation framework towards zero-carbon building performance. Especially for rapid construction activity in developing countries but also the existing building stock in developed countries, the future-proof design approach, supported by tools such as the DGNB framework for climate-neutral building stock or the individual renovation roadmap (iSFP), can be instrumental in overcoming inherent individual market barriers.

Author details

Dirk Schwede
Institute for Building Energetics, Thermotechnology and Energy Storage (IGTE), University of Stuttgart, Stuttgart, Germany

*Address all correspondence to: dirk.schwede@igte.uni-stuttgart.de

References

[1] Fisch MN, Wilken T, Stähr C. EnergiePlus Gebäude Und Quartiere Als Erneuerbare Energiequellen. 2012. p. 328. ISBN: 9-783000-391675

[2] Schwede D, Lu Y. Potentials for CO2-neutrality through energy-retrofit of the existing building stock in 26 cities in China. Procedia Engineering. 2017;**198**:313-320. DOI: 10.1016/j.proeng.2017.07.088

[3] Hermelink A, Schimschar S, Offermann M, John A, Reiser M, Pohl A, Grözinger J. Comprehensive Study of Building Energy Renovation Activities and the Uptake of Nearly Zero-Energy Buildings in the EU—Final Report. 2019. Available from: https://ec.europa.eu/energy/en/studies/comprehensive-study-building-energy-renovation-activities-and-uptake-nearly-zero-energy [Accessed: 04 February 2020]

[4] Schwede D. Entwicklung des Gebäudeenergiebedarfs und Einsparziele in Deutschland und in verschiedenen Ländern in Asien. In: Lozán JL, Breckle S-W, Grassl H, Kuttler W, Matzarakis A, editors. Warnsignal Klima: Die Städte. 2019. pp. 255-262. Available from: www.klima-warnsignale.uni-hamburg.de

[5] Schwede D. Climate-adapted material research for the socio-economic context of Vietnam (CAMaRSEC). In: International Conference on Building Materials, Hanoi; 31 October–2 November 2019. 2019

[6] Fisch N, Möws B, Zieger J. Solarstadt: Konzepte, Technologien, Projekte. Kohlhammer; 2001. ISBN: 3-17-015418-4

[7] Phaal R, Farrukh C, Probert D. Roadmapping for Strategy and Innovation Aligning Technology and Markets in a Dynamic World. University of Cambridge, Institute for Manufacturing; 2010. p. 240. ISBN: 978-1-902546-82-7

[8] Rockström J, Gaffney O, Rogelj J, Meinshausen M, Nakicenovic N, Schellnhuber HJ. A roadmap for rapid decarbonization. Science. 2017;**355**:1269-1271

[9] Tapia O. Technologietransfer und umgang mit technologie. In: TRIALOG Zeitschrift für Planen und Bauen in der Dritten Welt, TRIALOG 25. 1990. pp. 18-20

[10] Braune A, Geiselmann D, Lemaitre C, Oehler S. Framework for carbon-neutral buildings and sites. German Sustainable Building Council (DGNB); 2018

[11] DWD. Ortsgenaue Testreferenzjahre von Deutschland für mittlere und extreme Witterungsverhältnisse, Bundesamt für Bauwesen und Raumordnung (BBR) durchgeführt vom Deutschen Wetterdienst (DWD). 2016. Available from: https://www.bbsr.bund.de/BBSR/DE/FP/ZB/Auftragsforschung/5EnergieKlimaBauen/2013/testreferenzjahre/start-node.html [Accessed: 07 February 2020]

[12] BMWE. Handbuch für Energieberater Anleitung mit Tipps und Tricks zur Umsetzung, Bundesministerium für Wirtschaft und Energie (BMWE). 2017. Available from: https://www.bmwi.de/Redaktion/DE/Publikationen/Energie/handbuch-fuer-energieberater.html [Accessed: 10 February 2020]

[13] DOE. DOE Zero Energy Ready Home Program, US Department of Energy. 2020. Available from: https://www.energy.gov/eere/buildings/zero-energy-ready-homes [Accessed: 11 February 2020]

Chapter 4

Net-Zero Energy Buildings: Principles and Applications

Maher Shehadi

Abstract

Global warming and climate change are rising issues during the last couple of decades. With residential and commercial buildings being the largest energy consumers, sources are being depleted at a much faster pace in the recent decades. Recent statistics shows that 14% of humans are active participant to protect the environment with an additional 48% sympathetic but not active. In this chapter, net-zero energy buildings design tools and applications are presented that can help designers in the commercial and residential sectors design their buildings to be net-zero energy buildings. Case studies with benefits and challenges will be presented to illustrate the different designs to achieve a net-zero energy building (NZEB).

Keywords: energy consumption, energy conservation, net-zero energy, building performance, sustainable development, sustainable energy sources

1. Introduction

Global warming and climate change are rising issues during the last couple of decades. Buildings including commercial and residential ones are major contributors to energy consumption [1]. Energy consumption in buildings significantly increases on a yearly basis due to the increased human comfort needs and services [2]. Multiple factors affect the energy consumption used for cooling buildings such as wall structure, window to wall ratio, and building orientation in addition to weather conditions [3]. Energy consumed by buildings was reported to compose a relatively large proportion of the global energy consumption [4]. The building construction and the way it is operated and maintained have a significant impact on the total energy and water usage of the world resources [5].

Buildings are the primary energy consumers contributing to more than 40% of the US energy usage [6]. According to the US Department of Energy (DoE), the heating, ventilation, and air-conditioning (HVAC) systems consume approximately 17–20% of the total energy bill of any facility or building [7]. The world equipment demand for HVAC systems has increased worldwide from approximately 50 billion US dollars in 2004 to more than 90 billion US dollars in 2014 and for the United States from almost 11 billion to 19 billion US dollars over the same period [8].

Thermal characteristics of building envelopes have become of rising significance for designers and owners due to its relation to energy consumption reduction. Improper thermal insulations in buildings can lead to higher chances of surface condensation when air has relative humidity higher than 80% and when the convective and radiative heat transfer coefficients of the exterior walls are small [9].

 IntechOpen

The purpose of this chapter is to discuss benefits and design guidelines for zero energy buildings. NZEBs have tremendous potential to transform the way buildings use energy. In response to regulatory mandates, federal government agencies and many other state and local governments are beginning to move toward targets for NZEBs.

Many states in the United States are mandating many rules and regulations to reduce the buildings' energy consumption. For example, New York and California, which house more than 20% of the United States' population, produce less than 10% of its carbon emissions [10]. These two states are leading the way in decreasing energy use through the proliferation of net-zero energy building in addition to other strategies.

2. Building performance metrics

According to the US Department of Energy (DoE), a zero-energy building was defined as the building that produces enough renewable energy to meet its own annual energy consumption requirements [11]. According to the European Union Article 2, a nearly zero-energy building is a building that has a very high energy performance where low energy is required by the building which should be covered to a very significant extent from renewable sources including sources produced on-site or nearby [12].

There are several metrics that define the performance of buildings such as the net-zero site energy building, net-zero source energy buildings, net-zero energy cost building, and net-zero energy emission building.

The net-zero site energy building is defined as the building that produces as much energy as it consumes when measured at the site. The net-zero source energy building is the building that produces as much energy on an annual basis as it uses as compared to the energy content at the source. On the other hand, the net-zero energy cost building is the building that uses energy efficiency and renewable

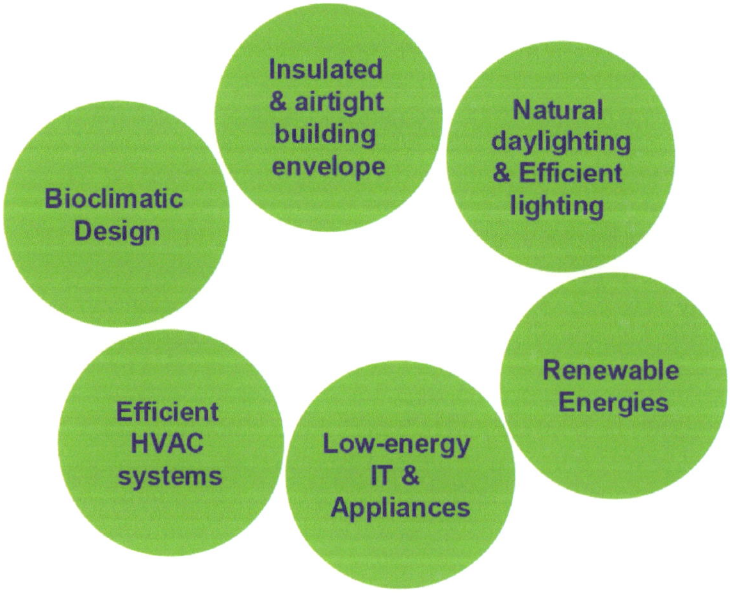

Figure 1.
Various energy efficiency measures.

energy strategies as part of the business model. Lastly, the net-zero energy emission buildings is the building design that looks at the emissions that were produced by the energy needs of the building. **Figure 1** shows various energy efficiency measures.

3. Why net-zero energy buildings?

In the last decade, energy costs have been rising, fuels are running out, and there have been global warming issues. For example, the United Kingdom has only 2 years of gas reserve, which has been put on hold of usage, and is currently buying from other countries such as Qatar and the United States. In addition to that, there have been many other issues such as health, well-being, and pollution which could be reduced if emissions are reduced as a result of better energy consumption plan.

Power stations convert only 30–35% of the input energy into electricity. The rest is rejected as waste heat. The United Kingdom alone wastes £20 billion each year by heat rejection from power plants which if used appropriately could heat Britain.

Earth's source of fossil fuel is vanishing at a much rapid pace during the last 200 years causing high damage rates to climate change. New reserves of fossil fuels are becoming harder to find. Those that are discovered are significantly smaller than the ones that have been found in the past. Oil reserve is expected to vanish between 2050 and 2060 and so does that for gas. Coal will last longer and is expected to last till 2100 [13].

Other aspects of increased emissions and increased rate of energy consumption are global warming and significant increase rate of ice melting and glaciers. A prominent red flag out of these aspects is that nine of the ten warmest years since 1880 have been in the last decade [14]. For global warming concern, Miami has seen a temperature rise of 3°C.

A building that is designed to be more sustainable has the potential to reduce the human impact on the environment. This effect is shown in **Figure 2**.

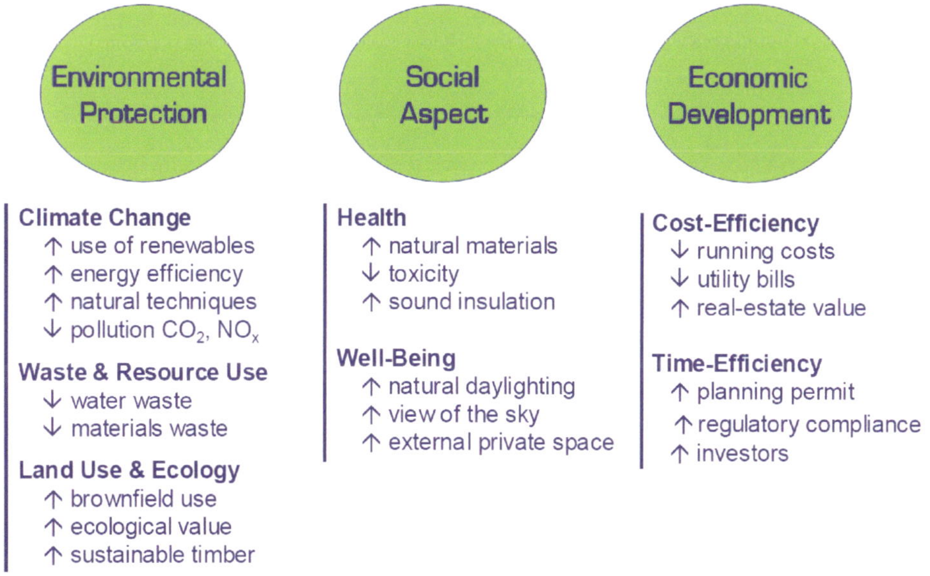

Figure 2.
Effect of sustainable buildings on the environment, social life, and economic development.

4. Sustainable development

Sustainable development is the development that meets the present needs without compromising the ability of future generations to meet their own needs [15].

There are three pillars for sustainable development:

i. Environmental protection

ii. Social concerns

iii. Economic development

The environmental protection aspect deals with climate change issues, resource depletion, land use and ecology, and waste concerns and impact of cities. The human social concerns and issues deal with justice, intragenerational equity, intergenerational equity, and health and well-being issues. On the other hand, the economic development deals with developed and developing counties, employment, modernization, and technological changes.

To solve current issues toward sustainable designs, designers should meet most of the items listed under each of the three pillars. These could be visualized as the intersection common areas shown in **Figure 3**.

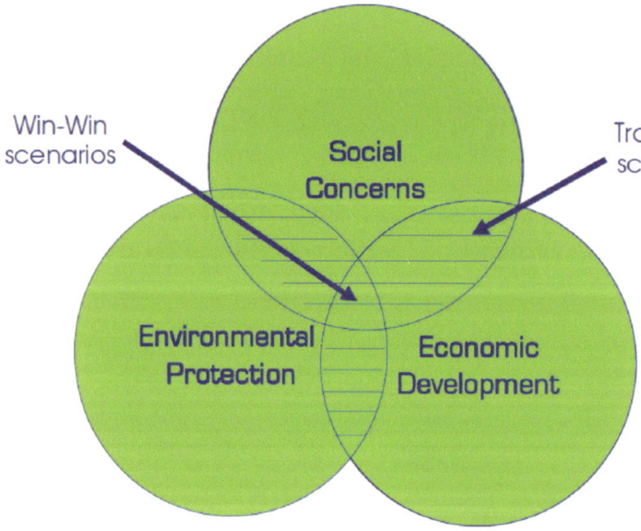

Figure 3.
Designers' choice to achieve the best results that meet sustainable designs.

5. Moving toward sustainable development and net-zero building designs: what it takes?

Spreading knowledge and engagement are ultimately the top most factors to help in reducing energy consumption, pollution and emission, and other issues such as global warming. The process starts with engagement and knowledge spreading, but it should be a closed cycle and thus needs feedback on performance. There has to be supplies that provide low and zero carbon energy and, lastly, investment. With no commitment from big industrial countries, no progress would be achieved.

There are many organizations who started net-zero marketing and application such as environmental organizations, research centers, universities and schools, and some engineering solutions which aimed to save costs and energy. In the United States, California and New York are leading the way to net-zero designs. Although they occupy more than 20% of the total population in the United States, they contribute to less than 10% of the total pollution emissions.

Following design standards is the first step in the design to achieve a net-zero energy building as it is important to define the sources and inputs that would be necessary to quantify the outputs and check what it needs to balance the net-energy consumed. The next step is to simulate the energy consumption using various energy modeling techniques and tools to optimize the following:

- Building orientation
- Glazing area, exposure, and shading
- Heat island reduction
- Lighting systems and capacities
- Temperatures, humidity, and relative humidity levels
- Landscaping
- Natural resources
- The overall system efficiency

All factors should be considered together by employing passive heating or cooling strategies, such as solar chimney and direct heat gain through south-facing glazing and/or isolated gain or sunspace, considering all possible exterior wall construction that avoids thermal bridging and increasing the R-value in all roof construction, using efficient lighting system, utilizing daylighting sensors and occupancy sensors, and lastly using energy-efficient office equipment for commercial buildings and energy-efficient utilities for residential houses and buildings.

The designer should then implement life cycle analysis, net-zero water system, and net-zero energy and optimize the design as per occupancy levels.

There are three principles to achieve a good net-zero energy building design:

A. Building envelope measures

Not only the building should be oriented to minimize HVAC loads, but shades and overhangs should be used to reduce the direct sunrays. Multiple options are available such as roof overhangs, shades and awning, and vegetation. To reduce the heat gain through windows, the designer should avoid glazing on the east/west façade. Other measures to reduce heat gains are to increase insulation on opaque surfaces, use glazing with low solar heat gain coefficient values, use double-skin façade, and refine the building envelope to suit location conditions.

B. Energy efficiency measures

The first utmost factor is selecting the right-size systems for the building. This can be achieved by following ASHRAE Standard 90.1 safety factors in the design, applying factors to reasonable baseline cases, and using simulation to model the design and predict the optimized requirements. In the simulation, part load performance should be considered which would come useful when using variable volume systems, variable speed drives, variable capacity boilers, variable capacity chiller systems, and variable capacity pumping systems as well. In addition to this, the designer should consider using high-efficiency lighting and control systems such as LED lights, high-performance ballasts, dual circuited task lighting, occupancy sensors, and daylighting dimming sensors.

The designer should shift electric loads during peak demand which would optimize the energy consumption. Some recommendations for optimizing the HVAC loads are (1) using heat recovery chillers, (2) using underfloor air distribution systems, (3) using high-efficiency chillers, (4) using passive cooling, (5) applying thermal storage using phase-change materials (PCMs), (6) using combined heating and power (CHP), and (7) using natural ventilation.

At the end of the construction phase, commissioning is a crucial step to ensure the building is performing as the intended design and is meeting its objectives. Commissioning phase verifies that the building's energy-related systems are installed and calibrated and perform according to the owner's project requirements, basis of design, and construction documents. The commissioning phase should cover at least the HVAC systems and controls, lighting and daylighting controls, domestic hot water system and any renewable system such as wind and solar. Building commissioning can reduce energy use, lower operating costs, reduce contractor callbacks, and improve occupant productivity. Successful implementation of the commissioning process can yield 5–10% improvements in the energy efficiency.

C. Renewable energy measures

Go green! Maximizing the energy sources are done through the first two measures, the building envelope which promotes using less energy and the efficient utilities and equipment measures. The renewable energy measures are more expensive than these two measures, and for that designers should start with the first two measures and optimize their design which would reduce the energy requirement needed in this step.

There are various renewable energy resources, such as solar which can be used for generating electricity, storing energy, and heating water, wind, biomass systems, and other sources.

Solar water heating systems include roof-mounted solar collectors that heat a fluid which would be used to heat water stored in a cylinder. Two collector types are usually used: the flat plate and the evacuated tube type. Flat plate collectors are usually cheaper. The solar water collectors heat the water that would be stored in a cylinder directly or indirectly by heating another fluid that would heat the water. Photovoltaic systems can be used to store energy and help in shifting the peak load.

Wind systems provide energy a very effective cost if the wind is continuous and steady and its speed above 10 mph (4.47 m/s), but it is recommended to be above 25 mph (11.2 m/s).

Biomass systems could provide heat by burning the biomass material. Some examples include forests, urban tree pruning, farmed wastes, wood chips, or pellets. However, the burners usually require more frequent cleaning than oil and gas boilers.

Geothermal systems provide good source for both cooling and heating by running the refrigerant pipes under the ground that usually provide nearly constant temperatures. These systems do not produce emissions. Such systems can provide coefficient of performance of 3 or even higher.

6. Applications with benefits and challenges

In this section, different case studies will be presented that implemented sustainable development and net-zero energy principles. The cases were selected based on their impact as reduced energy consumption and optimized sustainable resources used for energy and water.

6.1 The Bullitt Center

The Bullitt Center in Seattle was opened on Earth Day on April 22, 2013. The building is shown in **Figures 4** and **5** and is rated as the greenest commercial

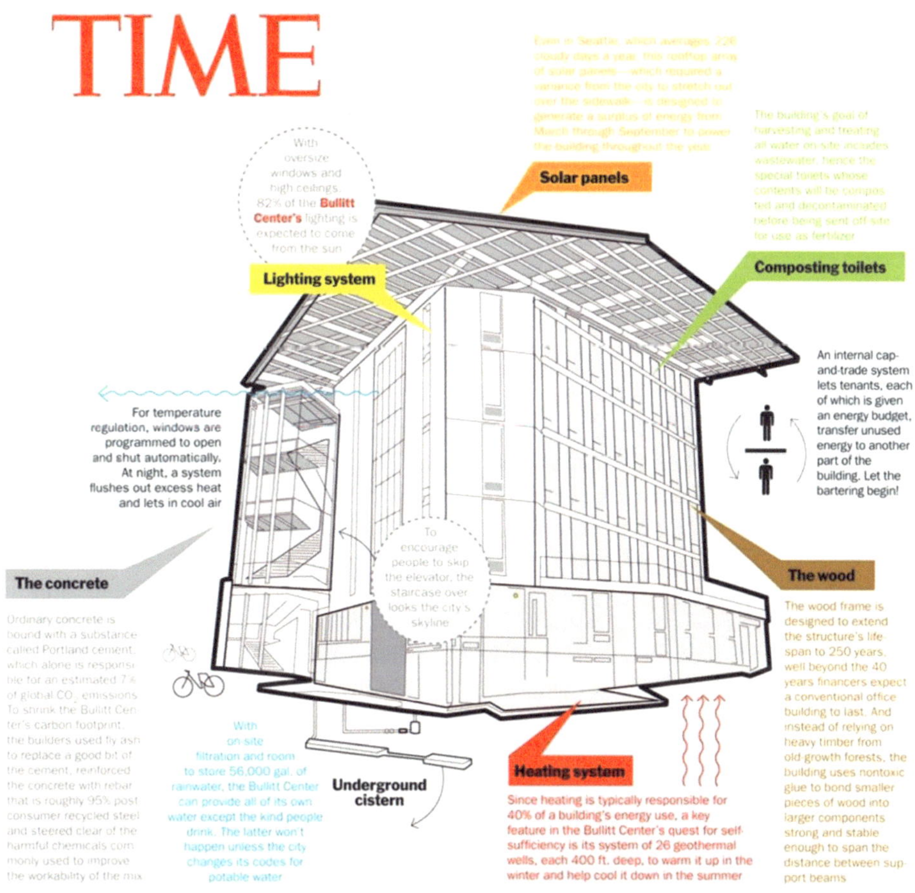

Figure 4.
Seattle's net-zero energy building (Bullitt Center) [17].

Figure 5.
Seattle's net-zero energy building (Bullitt Center).

building in the world. It is a six-story building and has a total area of 52,000 ft^2 (4800 m^2). The building is energy and carbon neutral, but its cost reaches as high as $18.5 million which yields $355 per square foot (per 0.09 m^2). The center's energy efficiency is 83% better than a typical office in Seattle with many efficient and sustainable energy sources including a 242 kW photovoltaic array, ground source geothermal heat exchange system, radiant floor heating and cooling, and retractable external blinds that block heat from warming the building. For water usage aspect, the center is 80% more efficient than a typical office in Seattle with live rainwater-to-portable water system that can collect up to 56,000 gallons (211,948 L) of rain-water [16]. The building also uses gray water reclamation using compositing foam flush toilets that save up to 96% of water as compared to traditional flush toilets. The building has also green roof and wetlands.

6.2 La Jolla Commons

La Jolla Commons II is a 13-story office at the University Town Center which is considered to be one of the largest NZEB in the United States. The building has a total area of 415,000 ft^2 (38,555 m^2) and was completed in April 2014 in San Diego, California. The completed building is shown in **Figure 6**. The building is rated as pre-certified silver as per US Green Building Council and a potential building for LEED platinum. The building has slab on-grade foundation. Other sustainability features include low-emissive coatings that reflect invisible long-wave infrared (IR) heat, reduce heat gain or loss in the building, and provide greater light transmissions. The walls were all glass as shown in **Figure 6** [18]. The air was supplied through underfloor air distribution (UFAD) system at 68 F (20°C). The cooling loads were 15 tons per floor and were supplied through two 560 tons cooling towers that served chillers located in the basement of the building. To achieve the net-zero energy efficiency, the building reduced the consumption through efficient designs and sustainable practices in addition to on-site generation. Fuel cells were generated at a rate of 5.4 megawatt-hour, whereas the historical expected consumption was approximately 4.5 megawatt-hour. The fuel cell technical data are shown in **Table 1**. The fuel cells are shown in **Figures 7** and **8**. The building is fed by biogas which would reduce energy costs. The cost per square footage was higher but it came with more benefits.

Figure 6.
La Jolla Commons [18].

Inputs	
Fuels	**Natural gas, directed biogas**
Input fuel pressure	15 psi, gage (6.89 kPa, gage)
Fuel required at the rated power	1.32 MMBtu/h of natural gas
Outputs	
Base load output (net AC)	200 kW
Electrical efficiency (LHV net AC)	>50%
Electrical connection	480 V at 60 Hz, three- or four-wire three-phase
Physical	
Weight	19.4 tons
Size	26' 5" × 8' 7" × 6' 9" (8 m × 2.6 m × 2 m)

Table 1.
Technical highlights for the La Jolla Commons fuel cells.

6.3 Aldo Leopold Legacy Center

It is classified as one of the greenest buildings on the planet as depicted by the US Green Building Council Prez [19]. The project consists of three one-story buildings. The project is located in Baraboo, WI, with cold and humid air conditions, with over 11,900 ft^2 area (1105 m^2). It has a platinum rating from the USGBC LEED-NC with net-zero energy rating. The first features of this project were the reduction in water consumption which reached up to 65% through the usage of waterless urinals, dual-flush toilets, and efficient faucets. The other features were the efficient irrigation features implemented using crushed gravels instead

Figure 7.
Fuel cells used at the La Jolla Commons building.

Figure 8.
Fuel cells used at the La Jolla Commons building.

of blacktop or concrete paving which increased the rainwater infiltration and helped in blending the developed areas with the surrounding landscape which eliminated the need for irrigation. The utmost feature for this project was the significant reduction in energy usage which reached to 70% less than a comparable conventional building by using 39.6 kW rooftop photovoltaic arrays that produces more than 110% of the project's annual electricity needs. A sketch for the design is shown in **Figure 9**, and a picture showing the installed cells on the roof is shown in **Figures 10** and **11**.

The buildings were oriented properly to have the maximum solar radiation source. Not only ground heat pumps were used as sources for heating and cooling, but Earth tubes were used to preheat and precool ventilation air, as well. Windows were utilized and properly oriented toward the south to get the maximum daylight that can reduce heat needs and lighting. The window area was maximized to optimize these two factors as shown in **Figure 12**.

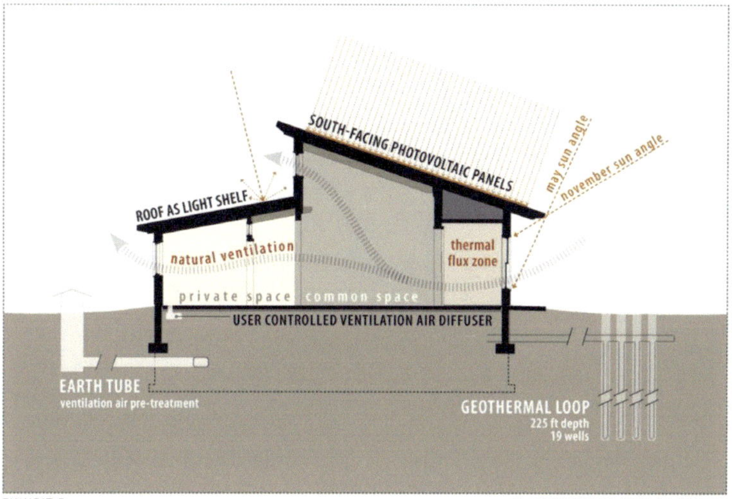

Figure 9.
Aldo Leopold Legacy Center in Wisconsin [20].

Figure 10.
Photovoltaic cells used for the Aldo Leopold Legacy Center project [20].

For additional heat, EPA-approved wood stove or fireplace was used. The final couple features were the usage of displacement ventilation and demand-controlled ventilation through the usage of variable frequency drives for fans that would control the amount of cooling or heating supplied to the spaces based on actual load and not the maximum designed.

The payback period for this project is expected to be around 14 years [20].

6.4 Hawaii Gateway Energy Center

The center is located on the island of Hawaii and is used by the Natural Energy Laboratory of Hawaii. The center is used for energy and technology research and development. The center is shown in **Figure 13**.

Figure 11.
Photovoltaic cells used for the Aldo Leopold Legacy Center project [21].

Figure 12.
Window orientation used to aid heating and lighting in the Aldo Leopold Legacy Center [21].

Figure 13.
Hawaii Gateway Energy Center [20].

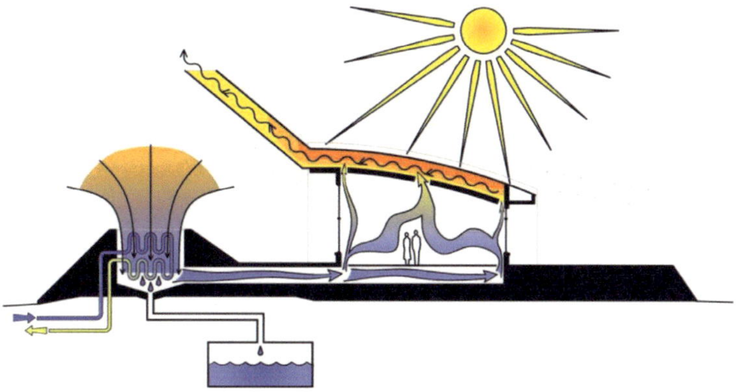

Figure 14.
Hawaii Gateway Center radiant roof system.

Natural ventilation is used through copper roof that radiates heat from the sun into a ceiling plenum as shown in **Figure 14**. Fresh outdoor air is pulled through the natural ventilation process into the occupied space from a vented underfloor plenum. Seawater at around 45 F (7.2°C) is used to cool the air to 72 F (22.2°C) as shown in **Figure 14**. As with the Leopold Legacy Center, the building is properly oriented to benefit from daylighting that aids lighting and reduces the energy needed to light the interior of the building. In summer, to prevent the negative affect of solar heat gain, shades are used on all windows. The center uses photo-electric daylight sensors to control the lights in addition to occupancy sensors. This prompted lights to be off 100% during daylight hours.

The building has 20-kW photovoltaic array which produces approximately 25,000 kW-hr due to high insolation in the area. Part of this power is used to power the pumps that draw seawater to cool the air and power the lights and other auxiliary electrical equipment. The building itself consumes 20% of the energy that comparable buildings use. In 2006, adjustments were made to the pumping systems which resulted in excess energy from the photoelectric system.

7. Conclusions

This chapter reviewed various techniques and designs that help achieve a net-zero energy building. The most important techniques are optimizing HVAC designs to reduce energy consumptions and usage of renewable sources. Some of the techniques include geothermal heat pumps, underfloor air distribution, radiant floor heating and cooling, retractable external blind on windows, and proper orientation of the building which would maximize heat gains in cold weather and minimize it in summer using trackable blinds, photoelectric daylight sensor, and occupancy sensor. Renewable sources include fuel and biomass cells, biogas, photovoltaic cells, and EPA wood stove for heating. Water usage as well could be optimized by using gray water reclamation and by using rainwater-to-potable live water systems.

Net-zero energy building design starts with ethical clients and demonstrators. Designers and users need to be lean in their designs to reduce the energy consumption, be clean by using energy-efficient utilities and systems, and be green by using renewable energy sources such as biomass, wind, solar, geothermal heat sink, and rivers. Canals could be a good source for heat pumps in cold weather regions [22].

Future buildings will focus more on renewable and sustainable energy resources by implementing an efficient building envelope and utilizing energy-efficient and high-performing utilities promoting reduced energy consumption levels. Future design will benefit from various potential energy resources including solar, wind, tidal, biomass, and other resources. Future system design and selection will need to simulate the various cases, variables, and scenarios to decide on optimized building design such as exposure, orientation, window to wall ratio, shading, building envelope, etc. In addition to that, artificial intelligence (AI) will play a major role in the operation and maintenance of such buildings including smart meters, smart display boards that recommend actions to tenants to reduce energy consumption, lighting control versus shading, and air-conditioning operation. Governments, local states, and cities have to commit to get this into track. They should facilitate sources access and should force using the guidelines and codes.

Nomenclature

NZEB	net-zero energy buildings
HVAC	heating, ventilation, and air-conditioning
DOE	Department of Energy
USGBC	US Green Building Council
LEED	Leadership in Energy and Environmental Design
LEED-NC	LEED-New Construction
EPA	Environmental Protection Agency
HEPA	high-efficiency particulate air
CFM	cubic feet per minute

Author details

Maher Shehadi
Purdue University, United States

*Address all correspondence to: mshehadi@purdue.edu

References

[1] Lin B, Li X. The effect of carbon tax on per capita CO_2 emissions. Energy Policy. 2011;**39**(9):5137-5146

[2] Shehadi M. Energy consumption optimization measures for buildings in the Midwest regions of USA. Buildings. 2018;**8**(12):170-186

[3] Dong H, Qi S, MA J, Liu S. Study on energy saving in residential buildings in the cold district. In: Proceedings of the 2010 International Conference on Advances in Energy Engineering. IEEE; 2010. pp. 351-354

[4] Li G. On the application of foamed composites in building energy efficiency. In: Proceedings of the Second International Conference on Mechanic Automation and Control Engineering. IEEE; 2011. pp. 6654-6656

[5] Gellings CW. Efficient use and conservation of energy in buildings. In: Clark W, Gellings PE, editors. Efficient Use and Conservation of Energy – Encyclopedia of Life Support Systems. Vol. 1. 2009. pp. 174-197

[6] EIA. U.S. Energy Information Administration. Available from: https://www.eia.gov/tools/faqs/faq.php?id=86&t=1 [Accessed: April 2019]

[7] Shehadi M. Office thermostat accuracy and reliability. Journal of Engineering Technology. 2019;**36**(1):16-27

[8] Mujahid RM, Gandhidasan P, Shafiqur R, Al-Hadhrami LM. A review on desiccant based evaporative cooling systems. Renewable and Sustainable Energy Reviews. 2015;**45**:145-159. DOI: 10.1016/j.rser.2015.01.051

[9] Sadineni SB, Madala S, Boehm RF. Passive building energy savings: A review of building envelope components. Renewable and Sustainable Energy Reviews. 2011;**15**:3617-3631

[10] Facilities Net. Net Zero Energy Buildings: How States Measure Up. 2020. Available from: https://www.facilitiesnet.com/energyefficiency/article/Net-Zero-Energy-Buildings-How-States-Measure-Up--17899

[11] US DoE. A common definition for zero energy buildings. Report prepared by The National Institute of Building Sciences for the U.S. Department of Energy; 2015

[12] European Commission. Energy. Available from: https://ec.europa.eu/energy/content/nzeb-24_en

[13] CIA. The World Factbook; 2020

[14] NASA. Climate 365. Accessed from: https://www.nasa.gov/content/climate-365-disagreement

[15] IISD. The Knowledge of Act. International Institute for Sustainable Development. Available from: https://www.iisd.org/topic/sustainable-development

[16] Bullitt Center. Rainwater-to-potable water system. 2019. Available from: https://bullittcenter.org/wp-content/uploads/2019/04/Bullitt-Center-Water-System-FINAL.pdf

[17] Solar Design Associates. Net-Zero Energy Bullitt Center. Available from: http://solardesign.com/portfolio/net-zero-energy-bullitt-center/

[18] Connors G, Panchuk I. La Jolla Commons II. 2013. Available from: http://www.josre.org/wp-content/uploads/2012/10/La-Jolla-Commons-II-by-Hines-by-Panchuk_Connors-4.30.13.pdf

[19] Alter L. Aldo Leopold Legacy Center: The "Greenest Building on the Planet." treehugger. 2007. Available from: https://www.treehugger.com/

sustainable-product-design/aldo-leopold-legacy-center-the-greenest-building-on-the-planet.html

[20] Solaripedia. 2009. Available from: https://www.solaripedia.com/13/103/924/aldo_leopold_hq_solar_panels.html

[21] The American Institute of Architects. Available from: http://www.aiatopten.org/node/135

[22] Department of Energy and Climate Change. Water source heat pumps – Navigating the way: A customer journey for potential developers Available from: http://www.gienergy.net/downloads/navigating_the_way_a_customer_journey_for_potential_developers.PDF

Section 2

Renewable Energy

Chapter 5

Solar Energy and Its Purpose in Net-Zero Energy Building

Mostafa Esmaeili Shayan

Abstract

The Net Zero Energy Building is generally described as an extremely energy-efficient building in which the residual electricity demand is provided by renewable energy. Solar power is also regarded to be the most readily available and usable form of renewable electricity produced at the building site. In contrast, energy conservation is viewed as an influential national for achieving a building's net zero energy status. This chapter aims to show the value of the synergy between energy conservation and solar energy transfer to NZEBs at the global and regional levels. To achieve these goals, both energy demand building and the potential supply of solar energy in buildings have been forecasted in various regions, climatic conditions, and types of buildings. Building energy consumption was evaluated based on a bottom-up energy model developed by 3CSEP and data inputs from the Bottom-Up Energy Analysis System (BUENAS) model under two scenarios of differing degrees of energy efficiency intention. The study results indicate that the acquisition of sustainable energy consumption is critical for solar-powered net zero energy buildings in various building styles and environments. The chapter calls for the value of government measures that incorporate energy conservation and renewable energy.

Keywords: net zero-energy building, electricity demand, renewable electricity, bottom-up energy analysis, solar energy

1. Introduction

The general description of a net zero energy building that can be found in the documentation is: "Net Zero Energy Building (NZEB) is a residential or industrial building with substantially decreased energy needs by productivity improvements, so that the balance of power requirements can be provided by sustainable technology." Even so, the researchers note that the "absolute zero energy structure" can be described in many forms, both on the parameter and the standard [1]. This description showed that energy conservation would be one of the techniques to achieve a net zero energy building output. The need for energy efficiency methods in NZEBs was already highlighted in several research sources. For example references of [2]. Focus were put on the preference for power conservation in the development of the NZEB and established the principle: 'first take up demand, then supply,' which implies that, in attempt to reach a net zero energy balance in buildings, it is important, first, to reduce power usage and power losses by energy conservation steps, lighting, ventilation, passive solar energy, high-efficiency appliances, thermal comfort, passive cooling, etc., instead and then using green electricity options to fulfill the energy requirements of buildings. Energy efficiency usually provides cost effective solutions for lowering electricity demand that significantly reduces the scale and thus the expense of the clean energy systems required and associated distributed technology [3].

This chapter aims to analyze synergy among power efficiency and on-site solar energy supply to move toward certain net zero energy quality. The results are taken based on the creation of the Building Integrated Solar Power System and the evidence from two other well-known field researches. The purpose of this study is to demonstrate that energy conservation and sustainable energy production are inseparable solutions and that the inadequacy of each of them has an enraging effect on the ability to reach a net zero energy target.

2. Research methods

Energy modeling is commonly used to estimate future energy consumption or the production of electricity in various industries. Two main types are generally reported in the literature: top-down and bottom-up [3]. Top-down methods utilize collated macroeconomic variables, including historical patterns, to create large-scale relationships between sectors in the economy [4]. The bottom-up studies depend on forecasts focusing on comprehensive technical and cost details from different sub-sectors, reflecting the total energy use of a nation or segment of the market [5]. Although bottom-up methods usually have even more comprehensive and consistent outcomes, the exposure and processing of disaggregated data needed for these models are sometimes tricky and often impossible.

Some studies develop the findings of a regional bottom-up approach that enables energy simulation of building energy usage and on-site solar energy generation with GIS techniques utilizing a variety of geographic information systems (GIS) techniques. GIS platform provides a broad range of methods for capturing, processing, extracting, and visually presenting geographically related results.

This modeling exercise's principal goal is to measure the full feasible hypothetical technological ability of building-integrated solar energy to satisfy the building energy needs and achieve a net zero value of building energy efficiency. The model assumes significant technical (and policy) advances to realize this solar energy promise by 2025. The simulation method, which is discussed in this article, consists of 3 key steps in tandem with various data sources. Although the author's BISE design is the key empirical tool for the findings provided in this section, the other two key components include only some of the data required to draw application provides and are thus defined in far less depth in this study. The mathematical descriptions of such models can be derived from the references seen in **Figures 2** and **3**. BISE method estimates the capacity for building-integrated solar energy supply, together with the findings on building energy usage from 3CSEP-HEB and BUENAS simulations, providing the ability to draw insights as to how much of these energy requirements can be fulfilled by solar energy in various regions and building styles.

2.1 CSEP: model of the HEB

3CSEP model of the HEB was established by the team of researchers (including the author of this chapter) at the Tarbiat Modares University to estimate the future usage of thermal energy building between 2015 and 2050 under a variety of policy-driven scenario. The design's central concept is a performance-based method for building energy consumption research, which views the buildings as a comprehensive structure rather than a collection of individual operating systems. In this method, the input variables of the main model has been the actual final energy efficiency of ideal houses (for each field, weather region, building size, vintage house) per square meter of its floor space obtained by the team of researchers from a variety of different sources recorded in [5]. Some rather building energy intensity levels are then compounded by the corresponding building floor space figures to measure the total energy usage independently for space heating, cooling and heat water in various countries, temperature zones, based treatment and vintages. That floor space has a different measurement formula for industrial and residential buildings that considers typical development activities such as relocation, reconstruction, and new growth, guided by demographic trends and economic growth shifts. This model integrates three scenarios, that imply specific levels of policy commitment in the area of energy performance construction and, accordingly, varying types of buildings energy efficiency in the national housing stock:

- The deep capacity paradigm presupposes an aggressive expansion of quality standards in energy conservation in buildings globally. Building energy efficiency is at the standard of passive design energy output (15–30 kilowatt hour/sqm for air conditioning based on the location).

- Strong performance scenario is poor continuity of current government patterns and small developments in energy quality construction in some developing countries. Built energy efficiency is at the standard of local building codes (100–200 kilowatt hour/sqm for air conditioning systems based on the location).

- Cold performance scenario suggests that the existing state of energy performance in buildings would stay constant throughout the studied span without implementing new policy tools or technical changes relevant to energy efficiency and conservation.

An in depth scenario has been used mainly to study the net zero energy building capacity because it implies substantial increase in power quality required to meet the NZE purpose. The effects of the energy usage from such a scenario are further compared to the projections of the BISE method's built in solar power capacity, as mentioned following.

2.2 Bottom-up energy analysis system (BUENAS) model

The BUENAS model presents the conclusion for energy usage in applications and illumination in the construction industry in order, which along with the findings of the heat energy use of such a 3CSEP-HEB method, render it possible to quantify the overall energy consumption in buildings.

- Lawrence Berkeley National Lab (LBNL) has established the BUENAS model for the end-use energy market scenario in the United States. This plan was sponsored by the Joint Marking of three Association Department.

This model approach produces outcomes for more than ten countries and the European Union with 27 members as a common area, including different energy-consuming goods (excluding appliances, such as TVs, laptops, etc.) in the domestic, commercial and industrial markets. The energy consumption prediction approach in BUENAS is focused on three main factors:

Two main scenarios mean the differences between the two models: Business as usual and best practice scenario. Under its scenario, energy consumption development is guided by market behavior and intensity. At the same time performance, is "frozen", the BP case focuses on catching future impacts of improvement-related policies, predicting that all governments can reach aggressive output goals by 2015. Standards will also have strengthened in 2020, ensuring whether the same degree of progress is attained in 2020 as in 2015 or which a particular goal, known as the new "best possible technology," is met by 2020.

2.3 Building integrated solar energy model

The author of this chapter, which considers different geographic, structural, morphological and climate conditions variables, has developed an alternative approaches Building Integrated Solar Energy model to assess the extent to which energy consumption could be met.

The BISE framework's primary goal is to analyze the highest allowable technical capacity and dynamics of solar power provided by built-in hybrid solar technologies. For this purpose, detailed climate data were taken from the NASA repository for some key variables (ambient temperature, top atmospheric irradiation, global irradiation, humidity data, wind speed, etc.).

The BISE method's additional advantage is exposure to high-resolution climate details, analyzing it via an advanced measurement method, extracting estimates for the future solar thermal and electrical performance of solar technology, and visualizing the estimates. This has been generalized for each area, outdoor environment, and site plan employing the roof area's various estimations to implement solar systems. The usable roof area is calculated by adding roof-to-floor ratios to the correlating floor area figures from the 3CSEP model as well as other access considerations extracted from the reference to compensate for the shaded areas and the gaps filled by roofing facilities. The RTR levels at each zone and buildings style are obtained by Geographic information systems datasets on regional urban development areas produced by Esmaeili Shayan [3] as well as further analyzed by the authors of this chapter utilizing Geographic information spatial analysis and zoning statistical techniques (see [3]). The spatial analysis's main objective is "to meet the demands and relationship issues, taking into consideration the spatial location of the phenomenon under investigation in a direct manner" [6].

While the roof area calculations primarily are using the floor area findings of the 3CSEP method as source evidence, the BISE method's configuration is quite close to that of the 3CSEP model in terms of areas, housing styles, temperature zones, and vintage architecture and simulation horizons. These elements of the layout are listed in more depth below.

2.3.1 Geographic coverage (GC)

In place to encourage a link between the results of the BISE and the 3CSEP designs, the analysis is carried out for whatever divisional division as presented in [7]. These areas of the country have included the following: Western Europe (WEU), Middle East (MEA), Centrally Organized Asia (CPA), Pacific OECD (PAO), Latin America and the Caribbean (LAC), Sun-Saharan Africa (AFR), former the Soviet Union (FSU), North America (NAM), Eastern Europe (EEA), South Asia (SAS), Other Pacific (PAS).

2.3.2 Timetable

The initial GIS data details were gathered for each hour of each year for 5 years from 2001 to 2005, and the 5-year estimate was determined for each level. Such data collection was used in the 2005 base year selected for compatibility with the 3CSEP model. The methodology considers every period and every year from 2005–2050.

2.3.3 End-use of energy

The Building Integrated Solar Energy model argument predicts that solar heat generated by PV/T systems can also be used for water and space heat generation. In contrast, solar electricity is used for lighting, space cooling, and appliances.

2.3.4 Climatic factors, construction vintages, housing styles

The Building Integrated Solar Energy model distinguishes between different types of buildings (residential: single- or multifamily; public and industrial: school, office, hotels and cafes, retail, health care, other housing, or buildings), vintages (retrofitting, modern, new, existing, and advanced retrofitting), seasonal conditions which are the same as the 3CSEP model.

2.3.5 Solar energy technology

The Building Integrated Solar Energy design focuses primarily on building-integrated on-site solar power. These systems can usually be broadly classified into two categories: solar thermal and photovoltaic (PV) systems. The latter produces heat, while the latter generates power. As the house needs both, maximizing the development of solar energy on the construction sites may demand the configuration of both kinds of processes. This might induce the "battle on the roof" (not enough space on the roof for both PV and solar collectors to meet energy demands) and lead to increased costs, esthetic problems, and a boost in the energy of the solar systems [3]. While solutions to this challenge currently exist by integrating solar systems with other innovations (e.g., photovoltaic + heat pump), since this chapter emphasizes exclusively on solar power, a thermal + photovoltaic hybrid solar system is perceived to be one of the most "fully solar" approaches to this problem. A solar hybrid photovoltaic/thermal system (PV/T system) is a mixture of photovoltaic (PV) panels and solar thermal elements. PV/T is a system that allows PV cells as a heated substrate to transform radiation into electric power; the solar thermal collector converts solar heat into electricity and removes waste heat from the PV module. These elements' goal is to use the heat produced in the PV panel to generate not only electrical but also thermal energy [8]. Such a hybrid setup generates an electrical utilization of the system as heat extraction and utilization reduces the systems' temperature and thus improve their performance. Configuration of photovoltaic plus thermal systems provides an opportunity to significantly increase the generation of solar energy for various end-uses compared to separate systems in the same roof area. As this chapter focuses on estimating the maximum possible technical potential of renewable energy in building structures, photovoltaic plus thermal technology was considered to be the most efficient model-long exercise workable alternative. In order to evaluate the hypothetical technological potential of built-in solar power, it is expected that photovoltaic plus thermal systems will be mounted on the available roof places during the construction or renovation of structures, beginning with some of those feasibility studies in 2014 then slowly expanding the number of installations before they become standard practice for all retrofits and housing developments by 2025.

The Building Integrated Solar Energy model assumes that thermal and electrical solar power production are modeled differently that use the same hourly in days' radiation exposure measured on 1 m^2 of the solar system site, but specific thermal and electrical formulations and performance variables and losses of different systems (see [3]).

2.3.6 Strategic partnership of electricity and energy performance in buildings

The Building Integrated Solar Energy model calculates the amount of solar renewable energy (electrical and thermal) produced in any buildings on an everyday hourly basis by BIPV/T systems, which is further compiled on a monthly basis. The present version of the product suggests the absorption of generated solar energy power within one period (month or more) at the rate of each city, buildings form, and temperature area, that makes it possible to equate the monthly amounts with the monthly projections of construction power consumption under the shallow scenario for space cooling, water heating and space heating and also with the Bottom-Up Energy Analysis System case formulation for home appliances. This scenario did not include industrial buildings. Consequently, the expectation that nearly 50 percent of cost savings attributed to energy efficiency changes in all end-uses should be reached by 2050 has been created. To achieve monthly results for equipment and lighting, it was presumed that these users would consume the same quantity of electricity every month. Monthly study results determined by the Building Integrated Solar Energy framework for the possible use of solar thermal energy have been evaluated by comparing to the construction energy consumption statistics for water heating and space heating. In contrast, the possible use of solar power for appliances and lighting, cooling was contrasted. Such a similarity forms the basis for assessing the extent to which advanced energy-efficient buildings with energy technologies can move toward the net-zero emissions energy systems target.

3. Calculation of the Shayan model

The novelty of the Building Integrated Solar Energy model integrates a comprehensive measurement process (acceptable for calculating the efficiency of the particular solar system) for hourly solar energy production per 1 m2 of the surface of the solar system and comprehensive coverage of the effects. The shayan model incorporates various forms of solar radiation, considering the tilting of the device (going to assume optimal tilting), the orientation of the earth, altitude, time of year, and location of the sun. The approach described here for measuring the energy obtained by one square meter of the solar system every hour has been modified from [1, 8–11]. There are many measurement benchmarks in the method. First, the total roof size was calculated in each area, outdoor environment, and building, which is mainly contributed by applying the accessibility variables.

3.1 Climate zone, building style, and area

$$AR = FR_{ratio} \times AF \quad (1)$$

where AR is area of roof and FR_{ratio} is floor of roof ratio and AF is area of floor. The Area Calculator (can be free use in: https://www.calculator.net) tools calculate

Figure 1.
The complex shape rooftop [3].

the area of the roof and the number of resources necessary to design the roof of the building. The "Home Foundation Field" is the land region that the building covers, which can be measured for more complicated forms using the Area Calculator. The measured area is an estimate only. In situations where a rooftop has a complex shape, such as **Figure 1**, calculating the measurements and areas of each part of the rooftop of a building to determine the total area would result in a more precise calculation of the surface.

3.2 Roof area for the integration of solar panels for each different environment region

Receiving energy from the sun is based on radiation. If the consumer is in the northern hemisphere, the sun's rays will be on the south side, and if objects are on the north side of the roof, they will cast shadows on the solar system. For various seasons, this impact would be different. Eq. 2 shows the space available for the use of the solar system.

$$AR_{Accessible} = F_s \times F_{rf} \times AR. \tag{2}$$

where $AR_{Accessible}$ is an accessible roof area for use in solar system installation and F_s is a factor of the impact of roofing facilities and F_{rf} is shading effect factor and AR is an area of roof accessible for use by solar systems.

Second, hourly solar radiation obtained by 1 m^2 of the solar system area is measured, considering the various forms of usable solar radiation.

3.3 Hourly solar energy on the plane

Complete radiation is one of the parameters for calculating the radiant energy of the sun. Global radiation depends on the variability of radiation and reflection in the environment. The following factors are: beam radiation, diffuse radiation, and reflective surface radiation. If I_{total} = full direct radiation from the atmosphere of the solar system and I_{global} = global radiation and ρ = the part of global solar radiation reflecting from the ground and $\left(\frac{1-\cos\alpha}{2}\right)$ is a factor of view to the ground and I_D = diffuse radiation and R_b is radiation ratio of the beam to the solar array on the flat plane, I_b, equal to beam radiation, then Eq. (3) can calculate the total radiation on beam.

$$I_{total} = I_{global}\rho\left(\frac{1-\cos\alpha}{2}\right) + I_D\left(\frac{1+\cos\alpha}{2}\right) + R_b I_b \tag{3}$$

The installation location of the solar system and the solar angles can affect the performance of the system. If this angle deviates from the vertical, the intensity of the radiation will also decrease. According to the definition, $\left(\frac{1+\cos\alpha}{2}\right)$ is a view element to the sun, that is, the proportion of the sun visible from the observation point (surface of the solar array) [3]. This variable can then be used to determine the thermal and electrical solar energy production independently by one square meter of a solar energy system per hour, bearing in mind the properties of the solar energy system and the ambient temperature, system errors, etc. Typical calculations for the electrical and thermal performance of individual solar energy systems have been used to achieve these tests. After this, solar electrical and thermal outputs are defined per square. The meter multiplies the estimates for the accessible roof area. Solar energy systems are configured based on geographical area, climatic zone, and type of construction. The hourly data for solar supply is then combined within each month of the year, implying the potential of solar storage systems within one month, and contrasted to the monthly predictions for building energy systems use for each month. Final uses (solar energy thermal production is compared with tests for room and water heating, solar energy electrical performance for ventilation, lighting, and appliances). This same full methodology of the BISE method is considerably further complicated and requires the further calculation of a variety of parameters described in this article. For further information, access to [12] is suggested.

4. Results

In order to emphasize the value of energy conservation for solar-powered NZEBs under the BISE model, the results for solar energy balances (i.e., solar energy supply vs. any building energy use) were compared to two 3CSEP scenarios: Deep conservation and medium efficiency categories for each of 11 countries, temperature areas, and based treatment. The essential purpose of such a study is to evaluate the effect on the solar fraction of the energy efficiency level change (i.e., the portion of building energy consumption that can be offset by solar energy output) in various regions and buildings. As noted above, extreme scenario presupposes very ambitious changes in energy quality (Approximately passive household energy efficiency), while moderate scenario assumes standard building energy output that can be attained by 2050 if existing government patterns proceed without significant innovations modifications.

The deep scenario results were combined with the energy use estimates of the appliances and lighting from the BUENAS model's BAU scenario with a 50% reduction in their energy intensities by 2050 to illustrate potential improvements in energy efficiency from these end-uses.

The result shows that the odds of meeting the net zero energy target in certain types of buildings are significantly smaller under the medium scenario than under the Extreme one. Tables also reveal that emerging regions can attain the NZE production over a more significant number of months than existing ones. The reason can be twofold: lower energy consumption in developing-country buildings due to more restricted access to modern energy infrastructure and a much greater abundance of solar energy supplies than in developed countries, most of which are concentrated in the northern hemisphere. This also demonstrates that in emerging regions (SAS, PAS, MEA, LAC, and AFR), the gap between the room and water heating energy usage is negligible. In these cases, electricity requirements for such end uses of most building styles (with some exceptions) in these regions can theoretically be fulfilled during the year by solar power supply only.

Full coverage can only be reached in other, primarily low-rise building forms (e.g., retail or single-family buildings) in all the months of 2050 in developing areas. The highest-rise structures, usually represented by multifamily and office buildings, display the lowest NZE capacity in developing regions among other building forms.

The number of months in which solar thermal is not adequate to satisfy the thermal energy demand in these buildings ranges from Low to high , depending on the location. PAO indicates the most significant potential for satisfying solar thermal energy demand across developing regions: Under the deep scenario, 100 percent thermal energy consumption coverage will be reached across all months and in all types of buildings. The great abundance of solar energy can explain this for most of the year in this area.

Results for the medium scenario explicitly demonstrate a substantial rise in the number of months, at least for developing countries, when thermal energy demands need additional energy sources and on-site solar power generation. Some of the situations in these countries, where a large amount of building energy consumption may be met with solar energy during the deep scenario for much of the months, would have some months in the medium scenario where it is not feasible. In the Medium case, only five building forms in PAO, single-family buildings in CPA, and residential buildings in WEU show the possibility in replacing thermal energy consumption with solar in all months.

Developing regions have ample solar power to meet solar heat thermal energy requirements during the year for most types of buildings, even with modest levels of energy efficient construction. In these countries, energy issues are still observed in some styles of tall and modern buildings. This is difficult to achieve monthly zero-energy ratios during the year (e.g., office and hospital buildings in SAS, PAS, and MEA, multifamily and school buildings in MEA).

As for electrical capacity, the disparity between scenarios in developed regions is more apparent—in the intermediate scenario, the number of months in which all electricity requirements can be met with solar energy than in the deep scenario in virtually all regions and building styles (exceptions are some categories of houses in the PAS and single-family homes throughout the LAC area, where maximum coverage age is possible in all cases during the month of the year).

Under the deep scenarios, emerging areas display a strong probability of supplying the bulk of building forms with ample solar electricity volumes. Nonetheless, the results for two high-rise building forms in MEA and office buildings in LAC indicate that solar power will not be adequate to satisfy the energy the building needs over the months. The mixture of thermal and electrical results provides an understanding in which regions and forms of buildings NZE efficiency can only be accomplished with solar energy based on the 2050 monthly energy balance. These cases would include:

- All styles of construction at PAS.
- The single-family PAO, SAS, EEU, CPA, MEA, LAC, and AFR buildings.

- SAS, LAC, AFR market constructions.
- The 'other' SAS, MEA, LAC, and AFR buildings.
- LAC educational institutions and hotels and restaurants AFR.
- LAC multifamily homes.

The findings set out in this document are predictions for (**Figure 2**) potential energy consumption in buildings by 2050 for different regions; building forms and end users and (**Figure 3**) the highest possible technological capacity for producing solar energy from advanced construction technologies.

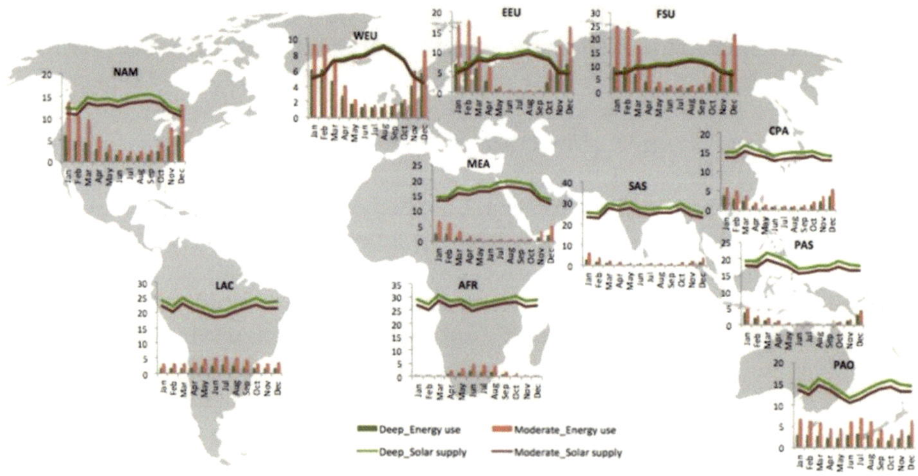

Figure 2.
For single-family buildings in 2050 in kilowatt hours per square meter of floor space, shown in deep versus medium conditions and the use of thermal energy versus solar thermal output [3].

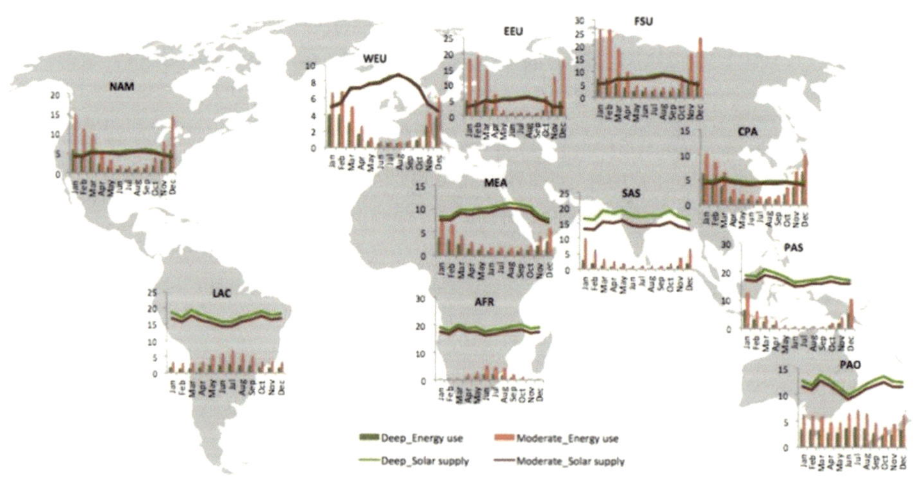

Figure 3.
Thermal energy usage vs solar thermal energy output in 2050 for industrial & public buildings, kWh / m2 of floor space, Extreme vs intermediate scenarios [3].

This chapter's key purpose was to compare the effects of building energy usage under two conditions with different levels of building energy efficiency to the amount of solar energy, which can theoretically be produced by advanced hybrid technology from the rooftops of these buildings. While solar energy capacity measurements have been conducted for each hour, the relation between solar energy supply and the building energy consumption is made monthly (due to the lack of more accurate statistics on building energy usage at the global and national level). It is estimated that generated solar energy will be accumulated at the construction site within 1 month at the level of each area, building type, and climate zone.

Five key messages can sum up the outcomes of such a comparison:

1. Synergies between energy conservation and on-site solar energy generation play a key role in bringing electricity output from building to net zero energy level.

2. Via "strong" energy conservation steps, the same volume of solar energy will support a more significant share of electricity demand, minimize the need for more energy from fossil fuels, and thus trigger greenhouse gases.

3. To exploit the net-zero energy performance capacity of all building services, including lighting and appliances, should be confirmed. New and updated buildings' thermal energy efficiency must meet passive house standards (about 15–30 kWh/m^2 depending on location and type of building). As for lighting and appliances, even halving their use of electricity by 2050 would not be enough to enable maximum solar coverage of the respective electricity needs in some regions (particularly developed ones).

4. Developing countries are seeing greater solar energy efficiency in buildings because of the availability of solar energy resources and lower electricity requirements. However, energy conservation is also critical in these regions to offset the substantial rise in energy usage anticipated in certain regions in the immediate future.

5. Low-rise buildings usually have a higher capacity to meet a significant portion of their solar energy requirement than high-rise ones. Yet modest energy efficiency standards make reaching the NZE target more difficult in most styles of buildings.

Author details

Mostafa Esmaeili Shayan
Tarbiat Modares University, Tehran, Iran

*Address all correspondence to: mostafa.esmaeili@modares.ac.ir

References

[1] M. Esmaeili shayan, G. Najafi, and A. ahmad Banakar, "Power Quality in Flexible Photovoltaic System on Curved Surfaces," J. Energy Plan. Policy Res., vol. 3, no. 17, pp. 105-136, 2017.

[2] R. Zeb, L. Salar, U. Awan, K. Zaman, and M. Shahbaz, "Causal links between renewable energy, environmental degradation and economic growth in selected SAARC countries: Progress towards green economy," Renew. Energy, vol. 71, pp. 123-132, Nov. 2014

[3] M. Esmaeili shayan, G. Najafi, and S. Gorjian, Design Principles and Applications of Solar Power Systems (In Persian), First Edit. Tehran: ACECR Publication- Amirkabir University of Technology Branch, 2020

[4] A. Walker, Solar Energy : Technologies and Project Delivery for Buildings. John Wiley & Sons, Inc, USA, pp. 153-192, 2013

[5] M. S. Esmaeili and G. Najafi, "Energy-Economic Optimization of Thin Layer Photovoltaic on Domes and Cylindrical Towers," Int. J. Smart Grid - ijSmartGrid, vol. 3, no. 2, pp. 84-91, Jun. 2019

[6] M. Azadbakht, E. Esmaeilzadeh, and M. Esmaeili-Shayan, "Energy consumption during impact cutting of canola stalk as a function of moisture content and cutting height," J. Saudi Soc. Agric. Sci., vol. 14, no. 2, pp. 147-152, Jun. 2015.

[7] M. Carmona and M. Palacio, "Thermal modelling of a flat plate solar collector with latent heat storage validated with experimental data in outdoor conditions," Sol. Energy, vol. 177, pp. 620-633, Jan. 2019

[8] M. Esmaeili-shayan, H. Shamsabadi, H. Azadi, Design and construction of a dynamic site to receive electricity from the river's flow studied by fishery and utilization workshops, in: Fourth Clean Energy Annu. Conf., Graduate School of Industrial and Advanced Technology, Iran, 2014

[9] M. Azadbakht, M. E. Shayan, and H. Jafari, "Investigation of Long Shaft Failure in John Deere 955 Grain Combine Harvester under Static Load," Univers. J. Agric. Res., vol. 1, no. 3, pp. 70-73, 2013.

[10] M. Azadbakht, M.E. Shayan, H. Jafari, E. Ghajarjazi, A. Kiapei, Factor Resistance Comparison of a Long Shaft in 955 and 1055 John Deere Grain Combine, International Journal of Mechanical, Aerospace, Industrial and Mechatronics Engineering, USA, (2015). doi:10.5281/ZENODO.1100909.

[11] J. D. Eichman, F. Mueller, B. Tarroja, L. S. Schell, and S. Samuelsen, "Exploration of the integration of renewable resources into California's electric system using the Holistic Grid Resource Integration and Deployment (HiGRID) tool," Energy, vol. 50, pp. 353-363, Feb. 2013

[12] G. Agarwal, A. Agarwal, and S. Agarwal, "BIPV : Integration of Photovoltaic with the Construction to achieve the concept of Agility Definition of BIPV". Asian Journal of Water, Environment and Pollution, IOS Press, Netherlands. 2014. http://ispe-usa.com/docs/images/icam2014_submission_35.pdf (accessed July 2, 2017).

Chapter 6

Computational Analysis of a Lecture Room Ventilation System

Abayomi Layeni, Collins Nwaokocha, Olalekan Olamide, Solomon Giwa, Samuel Tongo, Olawale Onabanjo, Taiwo Samuel, Olabode Olanipekun, Oluwasegun Alabi, Kasali Adedeji, Olusegun Samuel, Jagun Zaid Oluwadurotimi, Olaolu Folorunsho, Jacob Adebayo and Folashade Oniyide

Abstract

The level of Indoor Air Quality (IAQ) has become a big topic of research, and improving it using passive ventilation methods is imperative due to the cost saving potentials. Designing lecture buildings to use less energy or Zero Energy (ZE) has become more important, and analysing buildings before construction can save money in design changes. This research analyses the performance (thermal comfort [TC]) of a lecture room, investigate the use of passive ventilation methods and determine the energy-saving potential of the proposed passive ventilation method using Computational Fluid Dynamics (CFD). Results obtained showed that air change per hour at a wind velocity of 0.05 m/s was 3.10, which was below standards. Therefore, the lecture hall needs external passive ventilation systems (Solar Chimney [SC]) for improved indoor air quality at minimum cost. Also, it was observed that the proposed passive ventilation (SC) system with the size between 1 and 100 m^3, made an improvement upon the natural ventilation in the room. There was a 66.69% increase after 10 years in the saving of energy and cost using Solar Chimney as compared to Fans, which depicts that truly energy and cost were saved using passive ventilation systems rather than mechanical ventilation systems.

Keywords: computational fluid dynamics, indoor air quality, solar chimney, thermal comfort, zero energy

1. Introduction

Indoor Air Quality (IAQ) and energy consumption in lecture rooms is an important issue and of great concern during the last few years. For energy consumption, fossil fuels (coal, oil, and natural gas) are currently the most used form of energy resources, accounting for invariably 82% of energy consumed [1]. However, the burning of fossil fuels for our comfort emits carbon dioxide and in turn leads to the greenhouse effect, acid rain, and environment hazards [2–3]. Accordingly, a temperature analysis at NASA's Goddard Institute for Space Studies (GISS) noted that the average global Earth temperature has increased approximately by 0.8°C (1.4°F) since 1880, where most of the warming occurred in the last few decades at a rate of 0.2°C per decade since 1970 [4].

Another concern is the level of indoor air quality in lecture rooms which has been confirmed to cause discomfort among students in a class and may induce sleep and also affect learning. While considering the ventilation of lecture halls, indoor air quality is very vital. Natural ventilation has the potential to save energy costs as well as to maintain good air quality within the building, where natural ventilation is a method to deliver fresh air through buildings creating a pressure difference.

Natural ventilation could be as a result of air infiltration through various unintentional openings in the building. However, natural ventilation takes place as a result of manual control of opening of the building's doors, windows, or through other fenestration in the building. Furthermore, natural ventilation is achieved when a building is equipped with a ventilation system like natural ventilation solar chimneys and wind catchers. Air flows in and out of the building as a result of pressure differences across the openings, which is due to the combined action of wind and buoyancy-driven forces. In modern times, natural ventilation is considered not only as a simple measure of providing fresh air for the occupants and maintaining adequate indoor air quality levels, but also an excellent energy-saving means of reducing the internal cooling load of buildings located in the tropical regions. Natural ventilation system alone may achieve a good indoor thermal comfort, depending on the ambient conditions, without requiring the help of additional mechanical cooling devices.

1.1 Background of study

The high growth in population leading to high energy use globally and the accompanying increase in the fossil energy demand has resulted in detrimental effect on the environment and health of the society [5]. Among all users of fossil energy, the building industry sector has been observed to be one of the major energy consumers [6]. It is estimated that the building construction industry accounts for approximately 40% of the world's energy consumption. The industry is also accountable for the emission of pollutants of which about 70% of the total global emitted Sulphur oxides and 50% of emitted CO_2 are credited to it [7]. The total amount of energy consumed by the building industry sector are in four stages: stage one is during the process of making the materials; stage two is during transporting the materials, stage three is during the construction of the buildings and lastly during their lifetime of the buildings energy is required for operation and maintenance of the buildings [8].

However, it is essential to maintain good indoor air quality in buildings, which is achieved by providing sufficient ventilation to ensure the removal of stale air and the supply of fresh air for the occupants using several methods. These methods include mechanical ventilation (using fans and ducts to move huge volumes of air with or without heating/cooling the air); air-conditioning (in which the temperature and humidity of the air supplied through fans and ducts are fully controlled); and natural ventilation (which makes use of the naturally occurring driving forces of wind and buoyancy). A hybrid approach has also been used in practice, which combines both natural forces and mechanical devices, usually fans, to provide adequate ventilation [9–10].

The main disadvantages of air-conditioning and fans are cost (operation and maintenance costs) and space required to house the necessary equipment.

Consequently, several architects and building design engineers are giving more attention to natural ventilation or mixed-mode systems. Predictions of ventilation scenarios are more accurately modelled for mechanically driven systems as the designer knows the flow parameters in the different components making up the system [9, 11].

From the time when more powerful, inexpensive computers were available, an additional tool has become available to designers – Computational Fluid Dynamics (CFD). This method analyses the airflow in a room by sub-dividing the space in the room into small cells and solving the equations governing the airflow and temperature distribution in each cell. This method allows changes to the geometry and operating conditions be made more easily, offering a perfect tool for the investigation of various ventilation options an early stage in the design process [9].

1.2 Statement of the problem

The electricity consumed due to the extensive usage of air conditioning units during hot weather results in increased peak electricity demand during this season. This high electrical energy demand invariably increases the consumption of fossil fuel, leading to increased atmospheric pollution and finally climate change. Therefore, using renewable energy sources such as wind and solar energy as passive ventilation methods to provide adequate indoor air quality is vital. These methods will solve direct challenges associated with building high energy usage and indirectly pollution and climate change caused by numerous mechanical ventilation systems [12–13]. As with passive ventilation, fossil dependency is eradicated and therefore natural ventilation system is a better alternative solution for indoor air quality.

1.3 Aims and objectives

The aim of the study is to develop a computational model of a lecture room to perform ventilation analysis with and without incorporating passive ventilation, and thus investigating the energy saving potential and indoor air quality improvements of the room.

The objectives were:

I. To analyse a model lecture room's thermal comfort performance without mechanical or passive ventilation devices using CFD.

II. To investigate the use of passive ventilation system to improve the natural ventilation of the room.

III. Determine the energy saving potential of the proposed passive ventilation system.

1.4 Justification of study

There is huge dependence on electricity to run mechanical devices to provide ventilation and thermal comfort in buildings located in the tropics and in temperate regions during hot seasons. Commercial and residential buildings account for 40% of the world energy demand as well as 40–50% of the global carbon emissions [14]. More than 60% of the total energy consumption in buildings are attributed to Heating Ventilation and Air Conditioning (HVAC) systems [15]. This signifies a major opportunity for reducing the total global energy consumption and greenhouse gas emissions. Naturally ventilated building designs using devices such as wind catchers and solar chimneys are progressively being used for increasing fresh air flow and reducing energy consumption in buildings. Proper natural ventilation is of necessity owing to the high cost of maintenance of mechanical ventilation systems and relative increase in electrical energy cost needed to operate such machine. It is therefore justified to apply wind and solar energy, which are renewable energy sources, to provide adequate thermal comfort, indoor air quality and solve some challenges associated with mechanical ventilation systems.

2. Literature review

2.1 Ventilation

According to Awbi [16] ventilation is the "Replacement of polluted or stale indoor air by fresh or unpolluted air from outside." The main purpose of a ventilation system, mechanical or natural, in a building is providing acceptable thermal environment and indoor air quality for its inhabitants. In summary, ventilation in a building is designed to;

1. Deliver adequate air for breathing and removing CO_2. The conventional level for the maximum concentration of CO_2 within an occupied space is 5000 parts per million, or 0.5% by volume for an exposure of 8 h as recommended in **Table 1**.

2. Eliminate huge quantity of contaminants and airborne toxins.

3. Cool the building and its occupants in hot seasons [16–20].

2.1.1 Ventilation and thermal comfort

Ventilation systems are designed and incorporated in buildings to provide comfortable microclimates in the ventilated spaces [16]. The microclimate in this circumstance includes a thermal environment and air quality [21]. Therefore, in the design of natural ventilation systems these two factors (providing acceptable air quality and thermal comfort environment) are considered.

The human body's thermal balance are considered to be affected by four environmental factors (air temperature, mean radiant temperature, air velocity and

	Minimum ventilation required (litre/s per person)		
Activity	**0.1% CO_2**	**0.25% CO_2**	**0.5% CO_2**
At rest	5.7	1.8	0.085
Light work	8.6–18.5	2.7–5.9	1.3–2.8
Moderate work		5.9–9.1	2.8–4.2
Heavy work		9.1–11.8	4.2–5.5
Very heavy work		11.8–14.5	5.5–6.8

Table 1.
Ventilation rates required to limit CO_2 concentrations [18].

water vapour pressure in the air) [22] and three human factors (metabolism, activity and thermal insulation of clothing) [23].

2.1.2 The role of thermal comfort

There are some several reasons why thermal comfort is important in the design of buildings. These are:

- Human satisfaction is significantly affected by thermal comfort.
- There is a direct relationship between the energy consumption of a building to the temperature its occupants try to achieve in their accommodation.
- Occupants of buildings would continue to do everything possible to make themselves comfortable. Usually more energy is applied in achieving this and may be doing away with a planned low energy strategy [24].

2.1.3 Providing acceptable air quality

Ventilation in a building is designed to provide acceptable thermal environment in addition to suitable indoor air quality. However, the ventilation rate required to provide cooling and a satisfactory thermal environment is higher than that required to provide acceptable indoor air quality alone [25]. **Tables 1** and **2** show the required and recommended minimum ventilation rates necessary in buildings to provide satisfactory amount of oxygen for breathing, to dilute the metabolic CO_2 and to dilute odour. It is noted that the amount of ventilation required in a building is highly affected by the activity.

Figure 1 shows the ventilation rates for fresh air control, provided by the Chartered Institute of Building Services Engineers (CIBSE), the minimum amount of required airflow rate is also greatly affected by the amount of smoking, and by minimising smoking in the indoor area, the required airflow rate decreases as considerably.

2.2 Natural ventilation in educational buildings

Educational buildings are usually segmented into rooms with a number of people involved in various activities when in use. Thermal comfort in these buildings is important for the convenience of the occupants. Ventilation designs of educational buildings are based on factors such as the projected number of

Source	Purpose of ventilation	Minimum recommended value (1/s/p)
CIBSE (1989)	oxygen for breathing	0.3
HSE Guidance Note EH22 (1988)	oxygen for breathing	0.5
CIBSE (1986: B2–3)	dilution of metabolic CO_2	5
HSE Guidance Note EH22 (1988)	dilution of metabolic CO_2	2
CIBSE (1986: B2–3)	dilution of odour	8
HSE Guidance Note EH22 (1988)	dilution of odour	9

Table 2.
Recommended amount of air to provide acceptable indoor air quality.

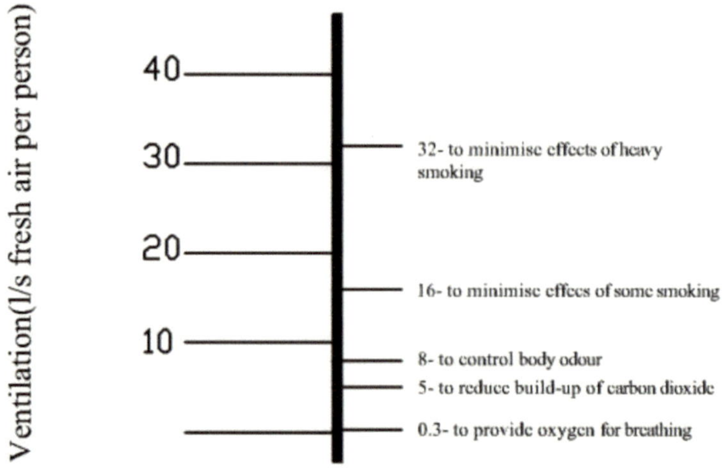

Figure 1.
Ventilation rates for fresh air control [26].

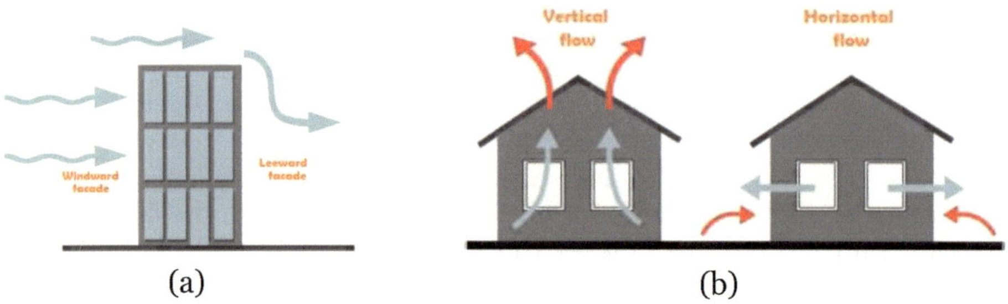

Figure 2.
Mode of natural ventilation: (a) natural ventilation due to pressure difference. (b) Natural ventilation due to temperature difference [27].

occupants and ambient conditions such as average temperature and relative humidity. In many places like Nigeria, power considerations require that energy consumption in these buildings should also be kept as low as possible. This may require the use of natural or passive ventilation which is strongly dependent on the airflow circulation pattern within room spaces, which can be achieved by temperature differences (buoyancy forces) or pressure differential between two points (**Figure 2**) [27]. Adequate ventilation in educational facilities is of great importance as reviews of previous studies in school environments [28–29] shows that there are inadequacies in the indoor air quality (IAQ) in many classrooms leading to higher risk of health-related issues especially in preschool environments. It was also noted that this problem is predominant in developed countries. However, the same challenges are encountered in developing countries especially during the hot season where classrooms are mostly overcrowded (**Figure 3**). In a study in China to improve the ventilation design of school buildings [30], it was noted that despite the fact that school buildings are actually designed for natural or mechanical ventilation, inadequate ventilation occurs. Furthermore, during hot season and due to crowded classrooms the ventilation system becomes inadequate thus affecting learning. It is therefore inferred that in tropical regions and developing countries where natural ventilation is being used, the number of students in a classroom, the outdoor temperature and the activities in the lecture hall affects the ventilation performance of the lecture hall.

Figure 3.
Selected (overcrowded) lecture hall.

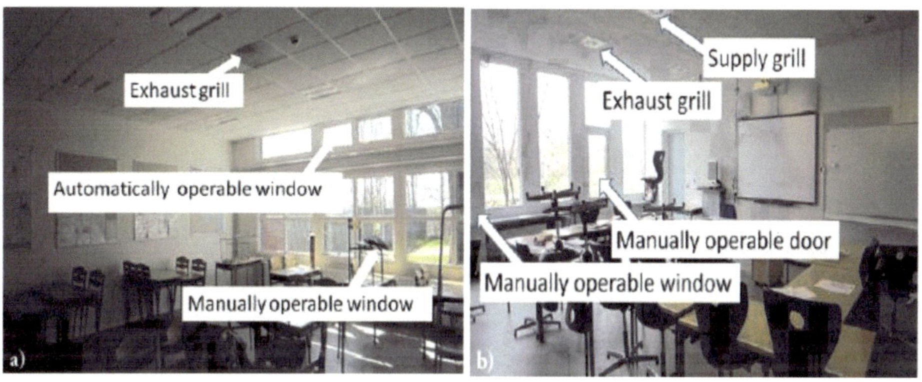

Figure 4.
Denmark study classrooms: (a) classroom with automatically operable windows and exhaust fan; (b) mechanically ventilated classroom [31].

In a study carried out in Denmark to measure indoor climatic conditions in classrooms [31], a comparison of different ventilation systems was done (**Figure 4**). The results obtained revealed that mechanical ventilation and natural ventilation with added exhaust fan performed better than the other systems. This indicates that the basic passive ventilation using window may not be adequate for educational buildings and may require the aid of other systems.

From other studies it has been observed that 30 students in a classroom would produce about 2.3 and 2.7 kWh of heat per hour and 500 litres of CO_2. These are indoor loads that need to be removed to improve the thermal comfort and IAQ of the classroom. Adequate classroom ventilation will solve these issues [32].

3. Methodology

3.1 Study area

The research was conducted at Olabisi Onabanjo University, College of Engineering and Environmental Studies, Ibogun Campus, Ifo, Ogun State, Nigeria located 240°SW on the Longitude 3.0990 and latitude 6.8080. The location has an annual average temperature of about 28.5°C and wind speed of about 4 m/s. The average relative humidity is about 63%. The study was performed in the selected lecture room where ventilation is normally achieved through opening of doors, windows and Mechanical ventilation system (ceiling fans). This lecture hall is representative of lecture halls in many Nigerian higher institutions especially in the southwestern region of Nigeria.

3.2 CFD model

The pseudo steady-state incompressible Reynolds-averaged Navier–Stokes (RANS) method is applied due to the computational cost and modelling accuracy. The standard $k - \varepsilon$ two-equation turbulence model has been modelled for this study. All room simulation scenarios modelled only considered wind-driven ventilation in isothermal conditions. The model equations are as follows.

3.2.1 Flow and energy equations

The general classical equations describing the flow in a room are represented as follows.

i. Continuity equation:

$$\frac{\partial P}{\partial t} + \frac{\partial}{\partial x_1}(pU_1) = 0 \tag{1}$$

ii. Momentum (Navier-Stokes) equation:

$$\frac{\partial}{\partial t}(\rho U_i) + \frac{\partial}{\partial X_i}(\rho U_i U_j) = -\frac{\partial p}{\partial X_i} + \frac{\partial}{\partial X_j}(-\rho\overline{u_i u_j}) + \frac{\partial}{\partial X_j}\left[\mu\left(\frac{\partial U_i}{\partial X_j} + \frac{\partial U_j}{\partial Xi}\right)\right] + g_i(\rho - \rho_r) \tag{2}$$

iii. Thermal energy equation:

$$\frac{\partial}{\partial t}(\rho T) + \frac{\partial}{\partial X_j}(\rho U_j T) = \frac{\partial}{\partial X_i}\left(-\rho\overline{u_i T'}\right) \tag{3}$$

It is noted that for the computational model used in this study that Eq. (3) has no effect on the airflow in the room as isothermal conditions were considered. However, Eq. (3) is considered for the solar chimney model. U_i represents the time-mean velocity component in the X_i direction while u is the fluctuating velocity components in the x_i direction. For turbulent flows, the viscous stress term on the right hand side of Eq. (2) is neglected as it is usually much smaller than the Reynolds stress term in the equation. For a case of steady incompressible flow and fluctuating velocities described by a suitable turbulence model, the effect of a fluctuating flow is represented by means of time-independent flow equations as shown below

$$\frac{\partial}{\partial X_j}(\rho U_i) = 0 \tag{4}$$

$$\frac{\partial}{\partial X_j}(\rho U_i U_j) = -\frac{\partial p}{\partial X_i} + \frac{\partial}{\partial X_j}(\rho\overline{u_i u_j}) + g_i(\rho - \rho_i) \tag{5}$$

$$\frac{\partial}{\partial X_j}(\rho U_j T) = \frac{\partial}{\partial X_i}(-\rho\overline{u_i T}) \tag{6}$$

3.2.2 Turbulence model

In Vector forms,
Turbulent kinetic energy (TKE, k)

$$\frac{\partial k\overline{u_i}}{\partial x_i} = \frac{\partial}{\partial x_i}\left(\frac{v_t}{\sigma_k}\frac{\partial k}{\partial xi}\right) + v\left(\frac{\partial \overline{u}_i}{\partial x_j} + \frac{\partial \overline{u}_j}{\partial x_i}\right) \tag{7}$$

Energy dissipation rate (ε)

$$\frac{\partial \varepsilon\overline{u_i}}{\partial x_i} = \frac{\partial}{\partial x_i}\left(\frac{v_t}{\sigma_k}\frac{\partial k}{\partial xi}\right) + c_{1\varepsilon}\frac{\varepsilon}{k}v\left(\frac{\partial \overline{u}_i}{\partial x_j} + \frac{\partial \overline{u}_j}{\partial x_i}\right)\frac{\partial u_j}{\partial x_i} - c_{2\varepsilon}\frac{\varepsilon^2}{k} \tag{8}$$

The solutions to Eqs. (5) and (6) require that the fluctuating velocity term in Eq. (5) and the fluctuating temperature term in Eq. (6) be represented by "equivalent" time-mean terms. All available turbulence models are semi-empirical and do not produce the same results. The two equation kinetic energy, k, and its dissipation rate, ε model is one of the most popularly used turbulence models applied by most researchers who studied the numerical solution of air flow in rooms and cavities [33–34] and is also used for the present study.

3.2.3 Setting up a CFD model

Computational fluid dynamics can be set up using the chart below in **Figure 5**.

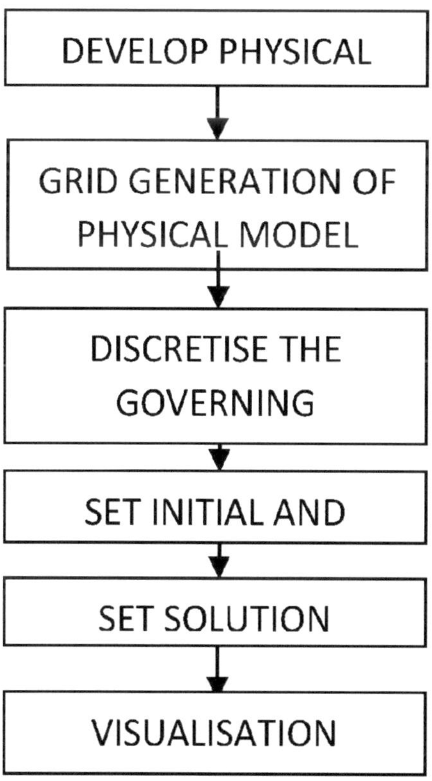

Figure 5.
CFD model chart.

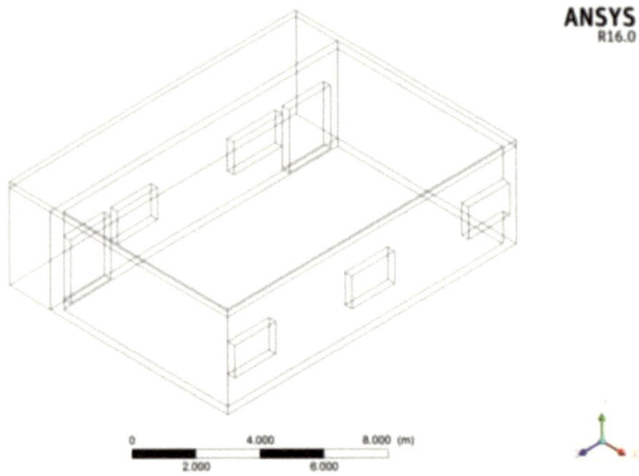

Figure 6.
Computational domain for analysis.

3.2.4 The physical model

This is the geometry of the area for the simulation. It constitutes the site plan and the room. The cross-ventilated building model used in this model is showed in **Figure 6**. The computational model was sized 17.4 m × 9 m × 3 m (length × width × height). The building height served as the reference length scale H. Two openings with dimension 2.4 m × 1.5 m (width × height) were installed at the rear end of both the windward walls.

The temperature range of the location is between 22 and 35°C, the outdoor relative humidity is between 35 and 90% and the wind speed is between 2.0–6.0 m/s. The lower value of the wind speed was used as the maximum for the computational analysis being the worst case scenario.

3.2.5 Grid generation of physical model (meshing)

In the CFD set up, the field is subdivided into several grids and the partial differential equations governing a flow field (e.g. velocities, temperature pressure, etc.) are solved at all points of the field [33, 35] as shown in **Table 3**.

In order to analyse fluid flow, flow domains are split into smaller sub domains. The process of obtaining an appropriate mesh is called grid generation (**Figure 7**).

3.2.6 Discretise the governing equations

After meshing, the governing equations are then discretised and solved in each of these sub domains. ANSYS Fluent uses the finite volume method for equation discretisation, which was used to perform the simulations in this study.

Number of nodes	Number of elements	Smoothing	Mesh type
2478	7337	Medium	Mixed • Triangular/Tetrahedron • Quadrilateral/Hexaheron

Table 3.
Grid analysis.

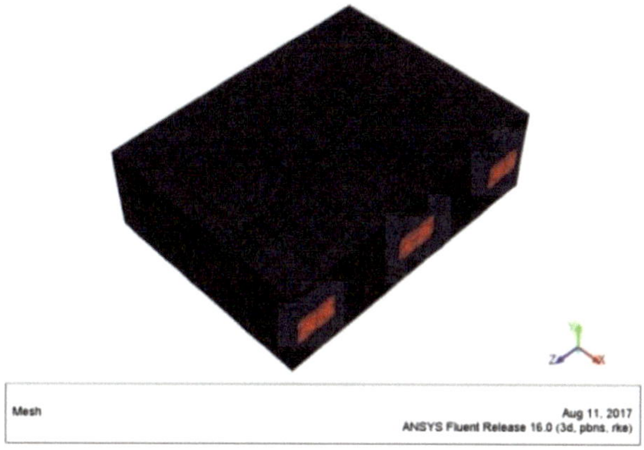

Figure 7.
Grid (mesh) generation for computational domain.

3.2.7 Initial and boundary conditions

The boundary conditions (**Table 4**) and the initial conditions are then set. The reasonable set up of physical quantities at the boundaries of the flow domain affect the overall accuracy of the CFD model. Initial conditions are essential for all CFD model. **Figure 8** indicates the point where solar radiation effect is introduced into the model, **Figure 9** shows the points initially set as flow outlets into the lecture hall, while **Figure 10** shows the wind inlet into the model. It is noted that an open corridor exists under the hanging roof and aid inflow to the lecture hall as shown in **Figures 8–10**.

3.2.8 Solution method

The solution method used is the SIMPLE Algorithm method. SIMPLE is an acronym for Semi-Implicit Method for Pressure Linked Equations. It is used to couple the pressure and velocity equations.

Wind velocity (m/s)	Solar radiation (W/m^2)	Pressure outlet (atm)	Turbulence model
0.05–2.0	200–1000	1	K - ε (Realisable)

Table 4.
Boundary conditions.

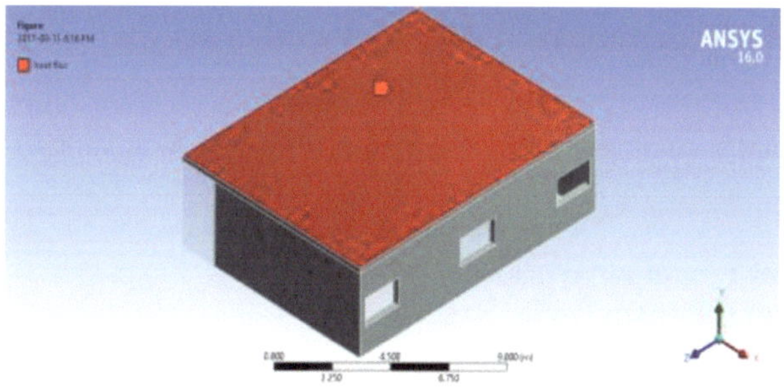

Figure 8.
Solar radiation input.

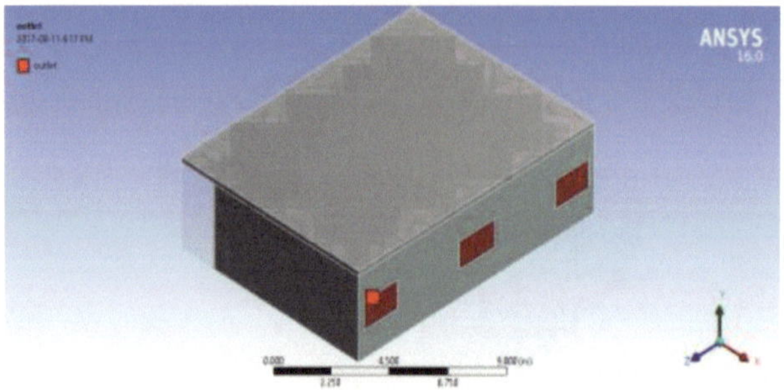

Figure 9.
Pressure outlet (mass flow) input.

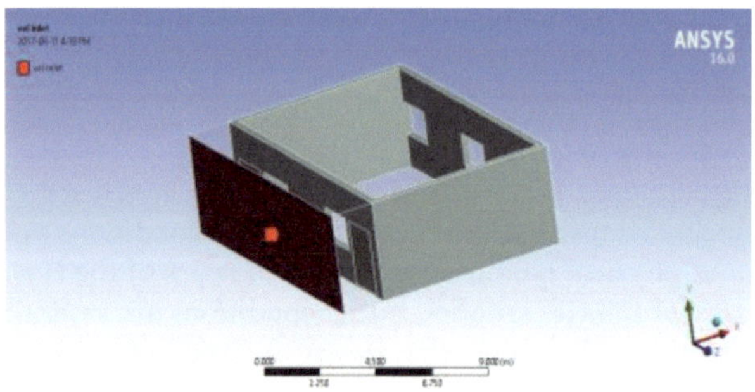

Figure 10.
Selection of the point for wind flow input to the model.

3.2.9 Visualisation

Results are the objectives of CFD simulation. They reveal the performance of the design. They reveal if the design satisfies or meet its objectives. Results are essential for making informed design decisions as revealed in **Figures 8–10**.

3.3 Performance analysis

Thermal comfort is determined by various factors. These factors include temperature, humidity and evaporative cooling. Using the values for the mass flow rate, the values for the air change per hour of the room can be deduced. Thermal comfort is very important to humans; therefore, using computational fluid dynamics analysis, ventilation performance of a room is determined with or without passive ventilation systems. The main driving force for the ventilation performance is the air change in the room depicted by the air change per hour, ACH.

The model for the performance analysis is written as:

$$ACH = \frac{3600m}{\rho f s V} \qquad (9)$$

where,
m = Mass flow rate of air (kg/s).
V = Volume of the room (m^3).

3.3.1 Solar chimney model

Using data from various researchers, values which include room size, solar chimney size, etc. have been used to investigate the use of passive ventilation

Room size (m^3)	Solar chimney size (m^3)	Ratio = SCS/RS	ACH (1/h)	Author
64	2.25	0.035	1.5	Bansal et al. [36]
50	2.5	0.05	3.46	Maerefat and Haghighi [37]
27	1.64	0.0607	5.6	Mathur et al. [38]
25	6	0.24	8	Khedari et al. [39]
15	2.5	0.167	7	Yan et al. [40]

Table 5.
Parameters of previous researches by various authors.

system (solar chimney) for improving the natural ventilation of a room. These data helped in developing a mathematical model that was used to predict the rate of ventilation (the air change per hour) of a room or space given by Eq. (10):

$$Y = mX + c \tag{10}$$

where,

$$X = \frac{Solar\ Chimney\ Size}{Room\ Size} \tag{10a}$$

and,

$$Y = \text{ACH (Air change per hour)} \tag{10b}$$

Since the various solar chimney models used had different sizes and different room sizes, a means to normalise the data was found by diving the solar chimney volume by the room size. All these data were presented in **Table 5**. This ratio of the solar chimney size to room size was then used to develop a mathematical model (Eq. (11)) to predict the ventilation rate of the system (**Figure 11**).

Figure 11 shows the trend-line graph of the solar chimney in which a model equation is obtained from a quadratic fit and given by;

$$y = -190.9x^2 + 77.753x + 0.0625 \tag{11}$$

$$R^2 = 0.8559.$$

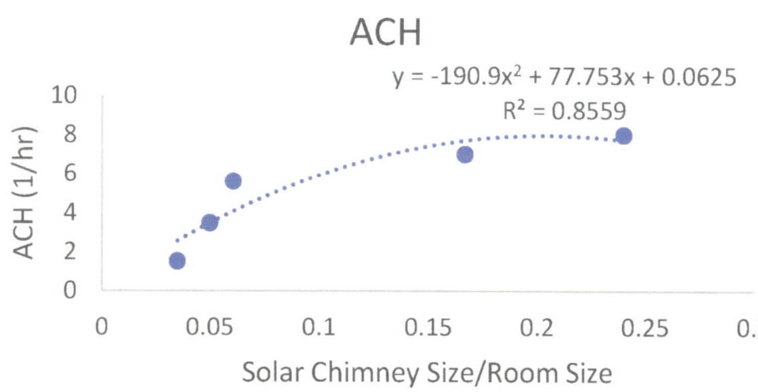

Figure 11.
Thread-line graph of solar chimney model.

Country [standard]	Outdoor air [m^3/h]	CO_2 concentration [ppm]	ACH [h^{-1}]
Portugal [RECS (2013)]	600	1250	4.0
UK [Building Bulletin 101 (2006)]	450	1500	3.0
Germany [DIN 1946–2 (2005)]	500	1500	3.3
USA [ASHRAE 62.1 (2013)]	558	1080	3.7
Europe [EN 15251 (2007)]	756	550	5.0

Table 6.
IAQ requirements in classrooms [41].

3.4 Energy saving model

This model was developed to reveal the energy saving potential of the passive ventilation system (the solar chimney). Energy saving is described as the level of energy saved and the reduction in cost derived from such saving. The model equations are as follows:

The cost of using mechanical ventilation system is given by,

$$\text{Total Cost} = \text{Installation Cost} + \text{Operation Cost} + \text{Maintenance Cost} \tag{12}$$

Also, the cost of using a passive ventilation system (solar chimney) is given by,

$$\text{Total Cost} = \text{Design Cost} + \text{Installation Cost} + \text{Maintenance Cost} \tag{13}$$

The percentage increase (energy saved) of both ventilation systems is given by,

$$\%Increase = \frac{C_{Mech} - C_{sc}}{C_{Mech}} * 100\% \tag{14}$$

where
C_{Mech} = mechanical ventilation cost.
C_{sc} = solar chimney cost.

3.5 Comparison against standards

Indoor air quality (IAQ) requirement is determined by various factors. **Table 6** shows some the acceptable IAQ international standards for classrooms. Therefore, the results obtained were compared with standard to validate the result.

4. Results and discussion

4.1 Air quality analysis

4.1.1 Lecture hall

Using CFD methods, the computational study was successfully performed to obtain the air properties for the lecture room. At the initial condition, where the wind velocity was 0.05 m/s, the air change per hour was found to be 3.1 and this value increases as the wind velocity increases.

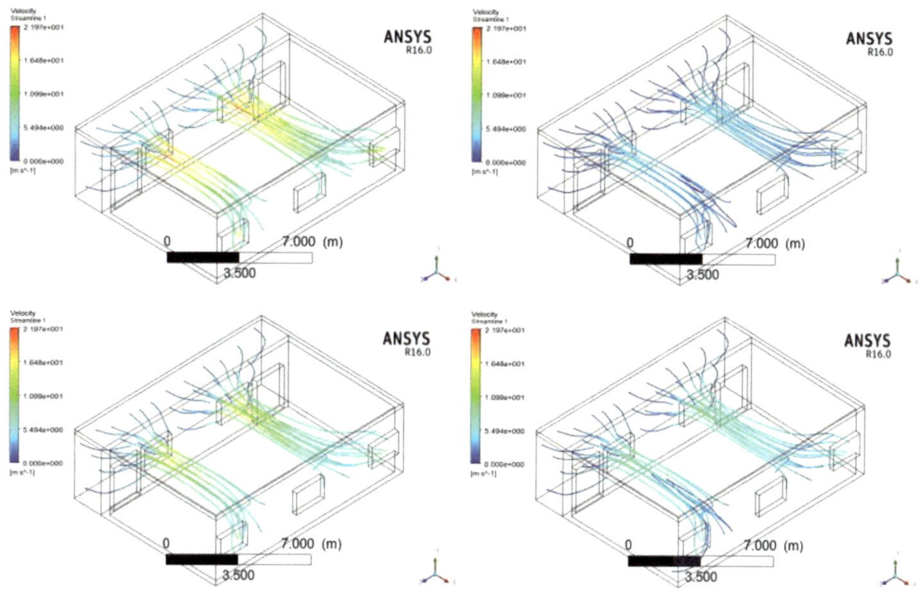

Figure 12.
Velocity streamlines of wind in the mechanical lecture room at different inlet velocities (0.05 TRH, 0.5 BRH, 1.0 BLH and 2.0 TRH m/s).

From the initial data, the results obtained fall short of the ASHRAE standard [42, 43], which proposed an average air change per hour of between 4 and 6 in a lecture room with an average wind velocity ranging from 0.12–0.5 m/s.

Figures 12–14 show the velocity streamlines, temperature contour, and velocity vector streamlines of the model.

4.1.2 Flow and thermal conditions of the lecture hall

Figures 12 and **13** reveal that the air velocities at the window sections are higher than in other places within the lecture hall. This is expected, as the airflow past the windows and enters the room, the flow velocity decreases as a result of the sudden expansion of the space in the room and increase in pressure according to Bernoulli's principle. It can also be observed that there are no obvious air movements at the centre and edges of the room. This is due to the initial condition that the inflow is from the side with two windows. This condition would not be pleasant for the occupants at these sections of the room. However, when the inflow comes from the sides with the three windows more sections of the room would have air considerable movement.

Figure 14 reveals that the air temperature distribution in the whole room is stratified. The lowest temperature appears to be at the side with three windows and highest at the areas with two windows. This is due to the higher convective heat transfer at the side with the three windows coupled with the increase in velocities at this section due to the venture effect. This indicates that areas closest to the three windows are more ventilated as compared to areas closest to that of two windows and this is a result of different in the number of windows.

Table 7 shows that the thermal comfort/air quality changes with air mass flow rate. As the mass flow rate increases air quality is also increased which shows that air quality is directly proportional to mass flow rate. At the initial state when the mass flow rate is 2.8 kg/s the air change per hour is 3.10. According to ASHRAE, in public places like the classroom, it must be ventilated in such a way that it has

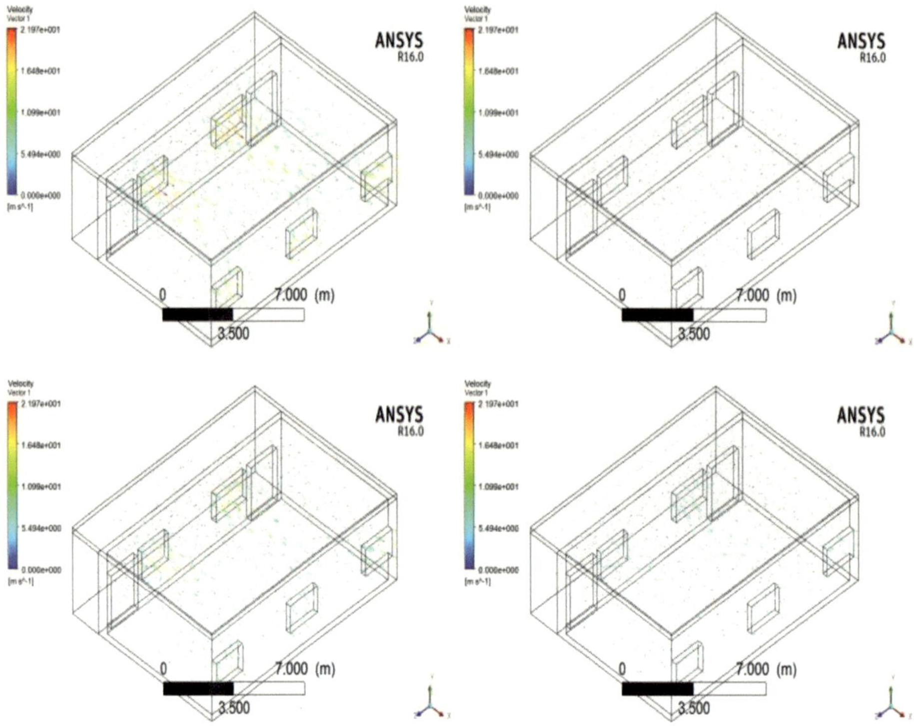

Figure 13.
Velocity vector of air in the mechanical lecture room at different inlet velocities (0.05 TRH, 0.5 BRH, 1.0 BLH and 2.0 TRH m/s).

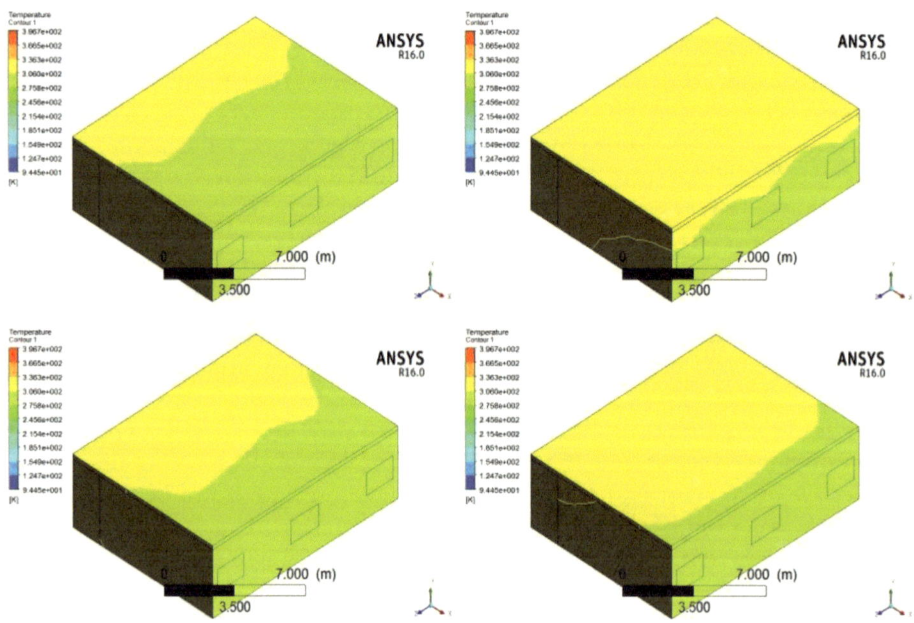

Figure 14.
Temperature contour of air in the mechanical lecture room.

Wind Velocity (m/s)	Mass Flow (kg/s)	ACH (1/h)
0.05	2.8	3.10
0.5	28.68	31.79
1.0	57.36	63.58
1.5	86.04	95.36
2.0	114.72	127.15

Table 7.
Mass flow rate and corresponding indoor air quality (ACH).

Airflow rate of 7.5 cfm/person. If the class will accommodate 100 students comfortably, the ventilated facilities must give 750 cfm air quality.

4.1.3 Room ventilation improvement methods

There are various means of improving the ventilation of the lecture room, they are:

4.1.3.1 Cross ventilation

It has been revealed by CFD simulations above that indeed the lecture room was inadequately ventilated at low wind speeds. The number/size of openings and introduction of another passive or mechanical system is required to improve the ventilation of the lecture hall. Positioning the ventilation openings such that a pair face each other to induce crossflows, and also positioning openings at the top allowing hot stale air to easily exit the building would create and optimise airflow and circulation pattern of the lecture hall. Windows or vents placed on opposite sides of the building induces natural flow pathway through the structure. Therefore, to improve the indoor air quality of this model, a cross ventilation of the lecture room and addition of topside opening would improve the thermal comfort and IAQ of the lecture hall.

4.1.3.2 Passive ventilation system (solar chimney) attachment

From the Eq. (11), a solar chimney model for the lecture room was developed and the graph shown in **Figure 11** was obtained. **Figure 15** reveals that using passive ventilation system such as a solar chimney with the right size, the indoor air quality of the lecture room can be improved. The graph is parabolic, which implies that some sizes of the solar chimney would work negatively when applied to the lecture hall. After the peak value (100 m^3), air change per hour begins to decrease with increasing chimney size.

4.2 Energy cost analysis

4.2.1 Energy saving potential

The energy saving potential of using a passive ventilation system was carried out in comparison to using mechanical ventilation system, which is given as follows:

4.2.2 Mechanical ventilation system's cost analysis

Installation Cost: The installation cost consists of the purchase cost and the fixing cost of the system i.e. fans. The average cost of purchase was found to be ₦5000 per fans and fixing cost was ₦200 per fans. For nine fans present in the room, we have:

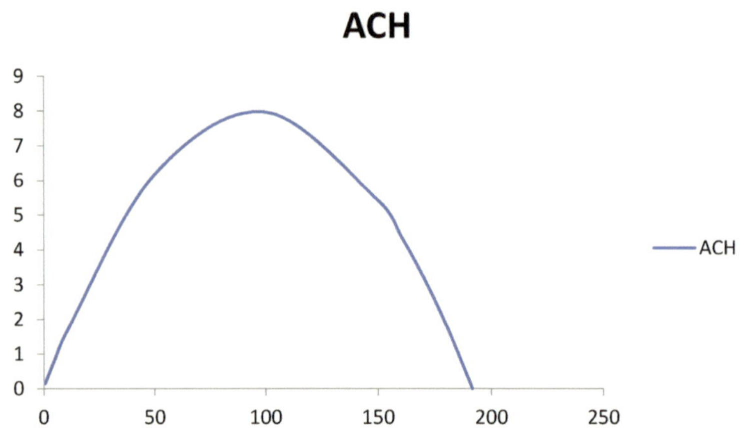

Figure 15.
Effect of solar chimney size on the air change per hour of the lecture room.

System	Unit Watt (W)	Number of fans	Energy demand (9 fans) (kW)	Hours/day (8 h)	kWh/ day	kWh/ year
Fan	25	9	0.225	8	1.8	657

Table 8.
Energy demand information of a unit of the ventilation device.

System	Cost (₦) per kWh	Cost/day (₦)	Cost/month (₦)	Cost/year (₦)
1 fan	25	5	150	1825
9 fans	225	45	1350	16,425

Table 9.
Annual energy cost for the mechanical ventilation system.

Installation Cost = (₦5000 * 9) + (₦200 * 9) = ₦ (45,000 + 1800) = ₦46,800.

Operation Cost: This is the average cost for the day to day running of the system. **Tables 8** and **9** shows the operational cost of running the fans.

Maintenance Cost: This is the cost of maintaining the system. It includes repair and basic cleaning. The average charge for maintenance was found to be at ₦500 per year.

Maintenance Cost = ₦500 * 9 = ₦4,500.

4.2.3 Passive ventilation system's (solar chimney) cost analysis

Design Cost: This entails the cost of materials used for the design e.g. wood, glass, pvc piping, washers, eye hook, grease, nozzle etc. and overall fabrication of the chimney. Using an exchange rate of ₦250 = $1.

Design Cost = ₦25,000.

Installation Cost: This includes cost of installing the design to the lecture room. Installation cost was found to be between ₦5000 and ₦15000. The average from this was obtained as ₦10,000.

Installation Cost = ₦10,000.

Maintenance Cost: It is referred to as the expense of using the solar chimney daily. The average maintenance cost was found to be ₦5000 yearly.

	DC (₦)	IC (₦)	OC (₦)	MC (₦)	TT (₦)
Mechanical System (Fans)		46,000	16,420	4500	66,920
Passive System (Solar Chimney)	25,000	10,000		5000	40,000

Table 10.
Total cost of ventilation systems.

Years	MVS (₦)	PVS (₦)	Increase (₦)	% Increase
1	66,920	40,000	26,920	40.23
2	87,840	45,000	42,840	48.77
3	108,760	50,000	58,760	54.03
4	129,680	55,000	74,680	57.59
5	150,600	60,000	90,600	60.16
6	171,520	65,000	106,520	62.10
7	192,440	70,000	122,440	63.63
8	213,360	75,000	138,360	64.85
9	234,280	80,000	154,280	65.85
10	255,200	85,000	170,200	66.69

Table 11.
Cumulative cost and percentage increase.

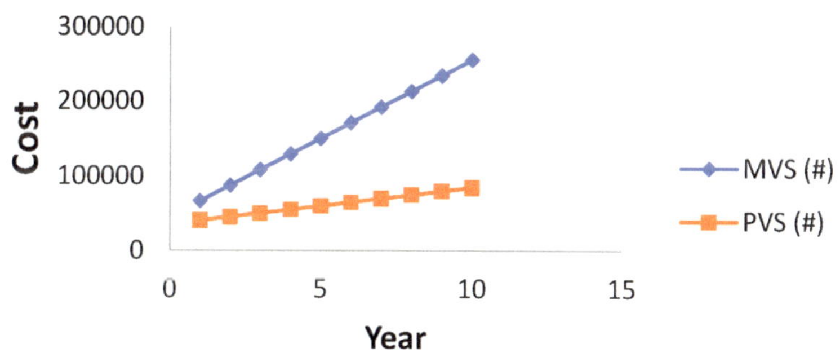

Figure 16.
Graph of the cost of solar chimney and fans against number of years.

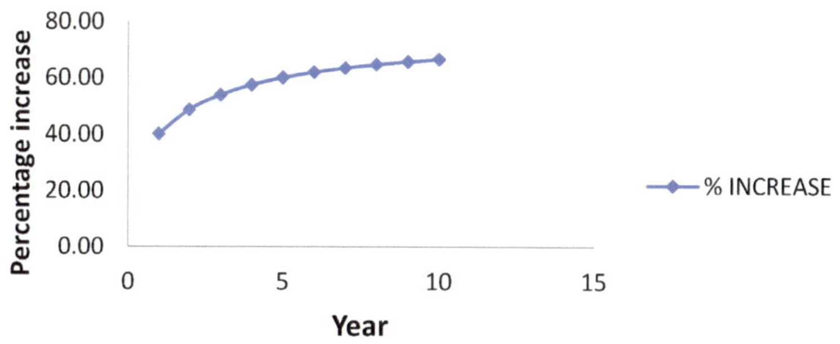

Figure 17.
Graph of percentage increase against number of years.

The total cost implications are presented in **Table 10**. The cumulative costs and respective percentage increases are presented in **Table 11**.

Figure 16 depicts the relationship between the cost of solar chimney and fans against number of years. It can be observed that the cumulative cost of running the mechanical ventilation system (fans) kept increasing continuously by over 30% each year and that of passive ventilation was lesser with about 12.5%. This indeed tells that the use of the solar chimney for ventilation is less costly, and invariably, energy efficient.

Figure 17 shows a relative increase in cost over number of years with the use of fans. It can be deduced that there was a 66.69% increase in cost due to the use of fans compared to the use of solar chimney in the lecture room for over 10 years. Therefore, with this percentage increase it can be understood that there is more energy and cost saving potential for the use of passive ventilated systems than mechanical ventilated systems.

5. Conclusion

Passive ventilation systems are a natural ventilation technique that has the potential of saving energy as well as maintaining good air quality in a building as compared to mechanical ventilated systems. This research was carried out with the aim of improving the ventilation performance (thermal comfort and indoor air quality) of a lecture hall. In this project, a detailed and systematic assessment of a lecture hall using CFD simulations was performed for determining the air flow caused by wind velocity stratification (0.05–2.0 m/s), solar radiation (200–1000 W/m^2) and pressure outlet of 1 atm within a space with seven ventilation openings. The air change per hour was deduced using different boundary conditions in the lecture room. The research revealed the airflow and temperature pattern within the room. It was obtained that a solar chimney of size between 1 and 100 m^3, improved the natural ventilation in the room. Also, from an energy and cost analysis carried out, cost-saving potential of the passive system (solar chimney) was established and found to be both energy and cost-effective in comparison to mechanical systems for better indoor air quality and thermal comfort. There was a 66.69% increase after 10 years in the saving of energy and cost using Solar Chimney (SC) as compared to fans. It is therefore acceptable to state that objectives were achieved and in turn a contribution to knowledge.

The knowledge of temperature stratification and mass flow rate is essential so as to aid the thermal comfort of occupants in indoor environments, especially in public buildings like lecture rooms, conference halls, etc.

Acknowledgements

The inputs and critique of members of the research team are acknowledged and appreciated. This research did not receive any specific grant from funding agencies in the public, commercial, or not-for-profit sectors.

Conflict of interest

The authors declare no conflict of interest.

Nomenclature

ACH	air change per hour
ASHRAE	American society of heating, refrigerating, and air-conditioning engineers
BC	before Christ
Btu	British thermal unit
CFD	computational fluid dynamics
cfm	cubic feet per minute
C_{mech}	mechanical ventilation cost
C_{sc}	solar chimney cost
CIBSE	chartered institute of building services engineers
CO_2	carbon dioxide
HVAC	heating, ventilation and air conditioning
HSE	health and safety executive
IAQ	indoor air quality
kg/m^3	kilogram per cubic metre
kg/s	kilogram per second
mm	millimetre
m^3	cubic metre
NASA	national aeronautics and space administration
RANS	Reynolds average Navier stokes
SC	solar chimney
TC	thermal comfort
UNCHS	United Nations centre for human settlements
^oC	degree celsius
ZE	zero energy

Author details

Abayomi Layeni[1], Collins Nwaokocha[1*], Olalekan Olamide[2], Solomon Giwa[1], Samuel Tongo[3], Olawale Onabanjo[1], Taiwo Samuel[4], Olabode Olanipekun[5], Oluwasegun Alabi[3], Kasali Adedeji[6], Olusegun Samuel[7], Jagun Zaid Oluwadurotimi[8], Olaolu Folorunsho[8], Jacob Adebayo[5] and Folashade Oniyide[1]

1 Department of Mechanical Engineering, Olabisi Onabanjo University, Ago-Iwoye, Nigeria

2 Department of Mechanical Engineering, Federal University of Agriculture, Abeokuta, Nigeria

3 Department of Architecture, Olabisi Onabanjo University, Ago-Iwoye, Nigeria

4 Department of Agricultural Engineering, Olabisi Onabanjo University, Ago-Iwoye, Nigeria

5 Department of Mechanical Engineering, University of Ibadan, Ibadan, Nigeria

6 Department of Mechanical Engineering, Lagos State University, Lagos, Nigeria

7 Department of Mechanical Engineering, Federal University of Petroleum, Effurun, Nigeria

8 Department of Computer Engineering, Olabisi Onabanjo University, Ago-Iwoye, Nigeria

*Address all correspondence to: yomilayeni@gmail.com; collinsnwaokocha@gmail.com

References

[1] U.S Energy Information Administration. Monthly Energy Review. 2015. pp. 1-214

[2] Pieprzyk B, Kortluke N, Hilje PR. The impact of fossil fuels - Greenhouse gas emissions, environmental consequences and socio-economic effects. In: Energy Research Architecture (ERA) Report. 2009. pp. 1-98

[3] Nwaokocha CN, Giwa SO, Kuye SI, Okezie SU, Adeala AA, Layeni AT, et al. Emissions characteristics of LPG retrofitted generator. Journal of Experimental Research. 2018;**6**(3):41-47

[4] Harsen J, Reudy R, Sato M, Lo K. Global surface temperature change. Reviews of Geophysics. 2010;**48**: RG4004

[5] Allard F, Chiaus C, editors. Natural Ventilation in Urban Environment: Assessment and Design. London: Earth scan; 2005

[6] Edwards B. Towards Sustainable Architecture: European Directives and Building Design. Oxford: Butterworth Architecture; 1995

[7] United Nations Centre for Human Settlements, (Habitat). Shelter Sector Performance Indication. Nairobi: UNCHS; 1993

[8] Asimakopoulos D, Santamouris M. Passive Cooling of Buildings. London: James and James science publishers; 2005

[9] Atiqur Rahman MD, Narahari GA. Analysis of lecture room using computational fluid dynamics natural and forced ventilation. International Journal of Engineering Research and Technology. Open Access. 2014;**3**(3): 653-659. ISSN: 2278-0181

[10] Layeni AT, Nwaokocha CN, Giwa SO, Sulaiman MA, Adedeji KA, Olanrewaju AI. Design and engineering economic analysis of a variable refrigerant flow (VRF) and mini-Split air conditioning system. Current Journal of Applied Science and Technology. 2019;**34**(1):1-25. DOI: 10.9734/CJAST/2019/v34i130111

[11] Layeni AT, Nwaokocha CN, Giwa SO, Olamide OO. Numerical and analytical Modelling of solar chimney for combined ventilation and power in buildings. Covenant Journal of Engineering Technology. 2018;**2**(1):36-58

[12] Liddament M. A Guide to Energy Efficient Ventilation, Coventry: Air Infiltration and Ventilation Centre. Sir William Lyons Road Coventry CV4 7EZ Great Britain: International Energy Agency, The Air Infiltration and Ventilation Centre, University of Warwick. Science Park Sovereign Court; 1996. pp. 1-274

[13] Nwaokocha CN, Alamu OJ, Adunola O, Giwa SO, Adeala AA. Performance evaluation of a domestic passive solar food dryer. In: Kale SA, editor. Renewable Energy System. New York, USA: Nova Science Publishers Inc.; 2017. pp. 67-84. ISBN: 978-1-53610-441-7

[14] Rahman MM, Rasul MG, Khan MM. Energy conservation measures in an institutional building in sub-tropical climate in Australia. Applied Energy. 2010;**87**:2994-3004

[15] Yau YH, Lee SK. Feasibility study of an ice slurry-cooling coil for HVAC and R systems in a tropical building. Applied Energy. 2010;**87**:2699-2711

[16] Awbi HB. Ventilation of Buildings. London: Spon; 1991

[17] Martin AJ. Control of Natural Ventilation (Technical Note TN 11/95). BSRIA: Bracknell; 1995

[18] Oughton DR, Hodkinson S. Faber and Kell's Heating and Air Conditioning of Buildings. Oxford: Butterworth-Heinemann; 2008. p. 2002

[19] Edwards B. Handbook of Domestic Ventilation. 10th ed. Oxford: Butterworth Heinemann; 2005. ISBN: 978-0-7506-8365-4

[20] Roaf S. Eco House. Oxford: Architectural Press; 2007

[21] Battle McCarthy Consulting Engineers. Wind Towers. Chichester: Academy; 1999

[22] Little J, Thomas R. Design with Energy: The Conservation and the Use of Energy in Buildings. Cambridge: Cambridge University Press; 1984

[23] Smith BJ, Philips GM, Smeeney M. Environmental Science. London: Longman; 1983

[24] Roaf S, Crichton D, Nicol F. Adapting Buildings and Cities for Climate Change: A 21st Century Survival Guide. Oxford: Architectural Press; 2004

[25] Heiselberg P. Principle of Hybrid Ventilation. Aalborg, Denmark: Aalborg University, Hybrid Ventilation Centre; 2002

[26] Building Research Establishment (BRE). Continuous mechanical ventilation in dwellings: Design, installation and operation. In: Building Research Establishment (BRE) Digest 398. Garston, Watford, WD2 7JR, United Kingdom: Building Research Establishment; 1994. pp. 1-8

[27] Oropeza-Perez I. Fundamentals of natural ventilation design within dwellings. In: Cakmakli A, editor. Chapter 3 in the Book "Different Strategies for Housing Design". Open Access, IntechOpen; 2019. DOI: 10.5772/intechopen.77454

[28] Salleh NM, Kamaruzzaman SN, Sulaiman R, Naziatul SM. Indoor air quality at school: Ventilation rates and it impacts towards children – A review. In: 2nd International Conference on Environmental Science and Technology IPCBEE. Vol. 6. Singapore: IACSIT Press; 2011

[29] Fisk WJ. The ventilation problem in schools: Literature review. Indoor Air, International Journal of Indoor Environment and Health. 2017;**27**(6): 1039-1051. DOI: 10.1111/ina.12403

[30] Songa J, Meng X. The improvement of ventilation Design in School Buildings Using CFD simulation. 9th international symposium on heating, ventilation and air conditioning (ISHVAC) and the 3rd international conference on building energy and environment (COBEE). Procedia Engineering. 2015;**121**(2015): 1475-1481

[31] Gao J, Pawel W, Wang Y. Thermal environment in classrooms with different ventilation systems. REHVA Journal. 2014;**51**(4):10-14

[32] Milles U. In: Milles U, editor. Ventilation in Schools: Better Learning Conditions for Young People. KaiserstraBe 185–197, 53113 Bonn, Germany: BINE Information Service; 2015. BINE-Themeninfo I/2015

[33] Layeni AT, Waheed MA, Adewumi BA, Bolaji BO, Nwaokocha CN, Giwa SO. Computational modelling and simulation of the feasibility of combined solar chimney for power generation and passive ventilation in buildings. Scientific African. 2020;**8**:e00298. DOI: 10.1016/j.sciaf.2020.e00298

[34] Awbi HB. Application of computational fluid dynamics in room ventilation. Building and Environment. 1989;**24**(1):73-84

[35] Etheridge D, Sandberg M. Building Ventilation: Theory and Measurement. Chichester: Wiley; 1996

[36] Bansal NK, Mathur R, Bhandari MS. Solar chimney for enhanced stack ventilation. Building and Environment. Great Britain. 1993;**28**(3):373-377

[37] Maerefat M, Haghighi AP. Passive cooling of buildings by using integrated earth to air heat exchanger and solar chimney. Renewable Energy. 2010;**35** (10):2316-2324

[38] Mathur J, Bansal NK, Mathur S, Jain M, Anupma. Experimental investigations on solar chimney for room ventilation. Solar Energy. 2006; **80**:927-935

[39] Khedari J, Boonsri B, Hirunlabh J. Ventilation impact of a solar chimney on indoor temperature fluctuation and air change in a school building. Energy and Buildings. 2000;**20**:89-93

[40] Yan MQ, Sherif SA, Kridli GT, Lee SS, Padki MM. Thermo-fluid analysis of solar chimneys. In: Industrial Applications of Fluid Mechanics. Proceedings of the 112th ASME Winter Annual Meeting, Atlanta, GA. 1991. pp. 125-130

[41] Almeida RMSF, Pinto M, Pinho PG, de Lemos LT. Natural ventilation and indoor air quality in educational buildings: Experimental assessment and improvement strategies. Energy Efficiency. 2017;**10**:839-854. DOI: 10.1007/s12053-016-9485-0

[42] ASHRAE. Approved American National Standard\American Society of Heating, Refrigerating and Air-Conditioning Engineers. Ventilation for Acceptable Indoor Air Quality Standard 62.1.-2007 (Supersedes ANSI/ASRAE Standard 62.1-2004). 1791 Tullie Circle NE, Atlanta, GA 30329: ASHRAE United States of America; 2007. ISSN: 1041-2336

[43] ANSI/ASHRAE, Approved American National Standard/American Society of Heating, Refrigerating and Air-Conditioning Engineers. Ventilation for Acceptable Indoor Air Quality Standard 62.1.-2013 (Supersedes ANSI/ASRAE Standard 62.1-2010). ASHRAE United States of America, 1791 Tullie Circle NE, Atlanta, GA 30329. 2013. ISSN: 1041-2336

Section 3

Sustainable Materials

Chapter 7

Fly Ash as a Cementitious Material for Concrete

Aissa Bouaissi, Long Yuan Li,
Mohd Mustafa Al Bakri Abdullah, Romisuhani Ahmad,
Rafiza Abdul Razak and Zarina Yahya

Abstract

This paper presents a review on fly ash as prime materials used for geopolymer. Due to its advantages of abundant resources, less in cost, great workability and high physical properties, fly ash leads to achieving high mechanical properties. Fly ash is considered as one of the largest generated industrial solid wastes or so-called industrial by-products, around the world particularly in China, India, and USA. The characteristics of fly ash allow it to be a geotechnical material to produce geopolymer cement or concrete as an alternative of ordinary Portland cement. Many efforts are made in this direction to formulate a suitable mix design of fly ash-based geopolymer by focusing on fly ash as the main prime material. The physical properties, chemical compositions, and chemical activation of fly ash are analyzed and evaluated in this review paper. Reference has been made to different ASTM, ACI standards, and other researches work in geopolymer area.

Keywords: fly ash, physical characteristics, specific gravity, particle shape, chemical activation, workability

1. Introduction

The production of ordinary Portland cement (OPC) contributes approximately 7% to the total of global greenhouse gas emissions; this is considered as a serious problem for the environment [1, 2]. Recently, much research has been done to explore an alternative product which could replace OPC. Geopolymer was firstly proposed by Davidovits [3] as an alternative binder to OPC. Geopolymer is defined as a chemical reaction of aluminosilicate compounds, which have geological origins, such as clay and metakaolin or from industrial by-products, for example, fly ash (FA) and ground-granulated blast furnace slag (GGBFS).

The collection of FA from the coal-fired power plants is based on different equipment and filtration processes. The source and the structure of burned coal significantly impacted on FA compounds. But, generally, all types of FA contain a huge amount of silicon dioxide (SiO_2) (both amorphous and crystalline), aluminum oxide (Al_2O_3), and calcium oxide (CaO) [4]. The typical chemical composition of FA which is obtained from Manjung power plant at Perak in Malaysia is presented in **Table 1**. It is classified as class F fly ash, which will be detailed in the next section. FA with a high percentage of silica (SiO_2) and alumina (Al_2O_3) (more than 80%)

Composition	SiO_2	Al_2O_3	Fe_2O_3	CaO	MgO	SO_3	K_2O	Na_2O
Concentration %	43.73	27.8	12.37	8.01	3.75	1.45	1.96	0.93

Table 1.
Chemical compositions of FA [5].

could be appropriate in the production of geopolymer as a raw material. According to Davidovits [3], geopolymers are classified as binder materials, which could be formed by the activation of aluminosilicate with alkaline solutions. The term "geopolymer" was first introduced by Davidovits and also well known as inorganic polymers or alkaline-activated binder material [4]. This review paper evaluates the significant characteristics of FA and its advantages as raw materials in geopolymer cement and concrete.

2. What is FA?

2.1 Classification

American Society for Testing and Materials (ASTM C618) [6] classified FA into two main classes based on the source of mineral coal; these categories are appropriately considered as important classes in the uses of concrete. The named class F and class C of FA have many similarities in terms of physical characteristics. However, a chemical composition analysis is required to distinguish between both classes. The total amount of silica (SiO_2), alumina (Al_2O_3), and iron oxide (Fe_2O_3) as the constituents of FA will determine the type of class. Fly ash is therefore classified as class F if the silica, alumina, and iron oxide content is at least 70% of the total mass and has a limited percentage of calcium oxide (CaO) (content no more than 10%). Class C FA constitutes at least 50% of silica (SiO_2), alumina (Al_2O_3), and iron oxide (Fe_2O_3) of the total mass and the calcium oxide (CaO) content is high (from 10 to 30%), with a high reactivity of almost all constituents [7].

Recently, many studies have been attempted in the analysis and synthesis of geopolymer. Some challenges have been faced in researching geopolymer process conditions and trying to identify the main aspects that limit and determine the reactivity of FA and geopolymers structure and its characteristics. FA-based geopolymer could be affected by many parameters [8, 9]; these parameters are significantly related to the primary materials and their characteristics, such as size and distribution of particles, the glassy phase in the content, the reactivity of both silicon and aluminum, constituent of iron, calcium and inert particles, and also the type of activator solution and its concentration.

Diaz et al. [10] supposed that the mechanical strength of geopolymer could be affected by many parameters of the mix design, for example, the ratio of NaOH to Na_2SiO_3 and activator solution to FA ratio. In addition, other factors could have significant impact on the behavior of fresh and hardened geopolymer, such as the physical and chemical properties and also the crystallographic of FA.

Particle size of FA could have a significant impact on the strength development in two ways. Firstly, when the particles are up of 45 μm, this has an influence on the water requirement in an adverse way. Particles size has an important effect on the reaction rate of FA at early stages. Secondly, once diffusion and dissolution of materials occur in concentrated pastes, surface area of the particles might play a considerable role in determining the kinetics of different processes [7]. Salloum [11] concluded that, from a study of 36 different concrete mixtures, there was a relationship linked to the fineness of FA and strength development in concrete.

Due to the fineness of ash particles, the reactivity level increases, this could appear in the case of low-calcium ashes compared to those of higher in calcium content.

2.2 Physical characteristics

The performance of concrete is significantly impacted by the physical characteristics of FA; these characteristics could be the volume, rheology, and water content in the slurry, pore distribution, and also the reactivity of constituents. **Table 2** presents different standards of pulverized FA (PFA) and its uses in concrete [7].

Brahammaji and Muthyalu [12] claimed that, the production of an optimal properties of a geopolymer binder, class F fly ash should contain less than 5% of unburned material, no high than 10% of Fe_2O_3 and lower in CaO content. Also the reactive silica amount should be between 40 and 50%, and 80 and 90% of particles should be smaller or in the range of 45 μm. A high amount of CaO leads to produce higher compressive strength, due to the formation of calcium-aluminate-hydrate (C-A-H) at the early age. The other characteristics which could influence the suitability of FA as a source material for geopolymers are, amorphous content, this means the amount of SiO_2, Al_2O_3 and Fe_2O_3 and also the morphology of FA. Other researchers [13] have reported that the amount of CaO + MgO could controls the characteristics of surface and the degree of progress of mortar and concrete carbonation. This occurs by providing anions and controls dosage requirements of water-reducing agents.

Sl no	Particulars	ASTM C618 Type F	BS 3892 Part 1	IS 3812
i	Particle density (kg/m^3, min)	Not specified	2000	Not specified
ii	Blaine fineness (m^2/kg)	Not specified	Not specified	320
iii	Retention on 45 μm (325 mesh) sieve (%, max)	34.0	12.0	34.0
iv	Loss on ignition (%, max)	6.0	7.0	5.0
v	Water requirement (% of PC, max)	105	95	Not specified
vi	Moisture content (%, max)	3.0	0.5	2.0
vii	Soundness (autoclave, max)	0.8%	10 mm	0.8%
viii	Strength activity index (%)[b]	75	80	80
ix	$SiO_2 + Al_2O_3 + Fe_2O_3$ (%, min)	70	Not specified	70
x	SiO_2 (%, min)	Not specified	Not specified	35.0
xi	Reactive silica (%, min)	Not specified	Not specified	20.0
xii	CaO (%, max)	Not specified[c]	10.0	Not specified
xiii	MgO (%, max)	Not specified	Not specified	5.0
xiv	SO_3 (%, max)	5.0	2.0	3.0
xv	Alkalis as Na_2O (%, max)[d]	1.5	Not specified	1.5
xvi	Total chlorides (%, max)	Not specified	0.10	0.05

[a]*The individual standards may be referred for more details.*
[b]*The 28-day compressive strength (N/mm^2) of blended cement mortar is expressed as the percent of that of the control Portland cement (PC) mortar. The ASTM standard for the purpose: ASTM C311,*
"Standard test methods for sampling and testing fly ash or natural pozzolans for use in Portland-cement concrete."
[c]*Not specified but generally below 10% when FA is produced from burning of anthracite or bituminous coal.*
[d]*The equivalent alkali content, expressed as Na_2O, is obtained as: $Na_2O + 0.658\ K_2O$.*

Table 2.
Comparison of some standards on PFA for use in concrete[a] [7].

2.2.1 Particle shape and form

Particle distribution and their size are considered the main physical factor for the geopolymerization process [14, 15]. Komljenovic et al. [16] stated that, the reactivity of FA increases with increasing its fineness, which leads to an improvement of geopolymer properties. Basically, the formation of ash particles occur during the condensation and liquefaction process of incombustible inorganic matter, which is remained after coal combustion [17–19]. The shape of FA particles depend on the combustion conditions and condensation process. In general, there are two major combustion processes. The first process occurs when the temperature ranges from 1204 to 1727°C, this process is called the pulverized coal firing system. The second process is known as fluidized bed combustion which could be peaked at temperature ranged between 827 and 927°C. Typically, the first process is the most common used one in the large thermal plants [20].

Surface tension of the melt plays a significant role in the formation of spheroidization of pulverized FA particles. Two types of particles could be formed, cenospheres, which are ash particles hollow from the inside, and plerospheres which are hollow ash particles but including smaller particles inside as is shown in **Figure 1**. Brouwers and Van Eijk [21] suggested that the formation of plerospheres is as a result of the cracking or puncturing of the primarily hollow particles during handling work, but not related to the melting process. Jayant reported that the shape and surface characterization of FA particles have an impact on concrete in terms of water demand, in particular at the desired slump stage [7]. The spherical forms of FA particles minimize interparticle friction and leads to the creation of a dynamic system between particles in a concrete. This process improves the flow properties of the concrete. An experimental study was carried out by Atiş et al. [22] on the properties of different types of FA. Their results showed that there are many similarities between the chemical and mineralogical composition of all types of FA and also the physical properties such as specific surface area, particle shape, and their distribution. To explain the performance of concrete from the strength and workability point of view, some authors proposed a new parameter called "shape factor" which is mainly based on the specific surface area of FA particles [7].

Another study shows that around 90% of tested FA could reduce water requirement of mortar mixtures. A correlation has been proposed to show the relationship between water demand and fineness and also water requirement and loss on ignition. Further, the addition of FA has a significant effect on the rheological properties of cement paste and workability of concrete, due to the small spherical particles of FA. Givi et al. [23] believed that the proportion of coarse material in the

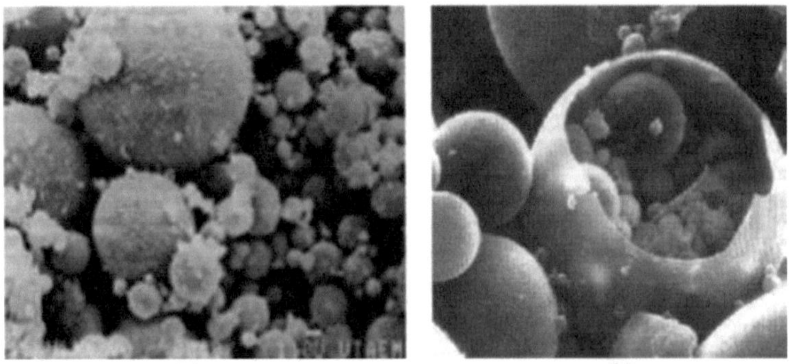

(a) (b)

Figure 1.
Scanning electron microscope of FA: (a) cenosphere and (b) plerosphere particles [7].

ash usually (up to 45 μm) is mostly the main parameter affecting the workability of concrete. A study carried out by Feng and Clark [24] confirmed that the water requirement has been effected by both sieved residue and loss on ignition (LOI), where the LOI has impacted on water demand, due to the absorption of water molecules by porous carbon particles.

2.2.2 Particle-specific gravity

According to ASTM C188 [25], the specific gravity of FA particles can be determined by the same method that is used for hydraulic cement. If there is a water-soluble molecule in FA, it is recommended to use nonaqueous solvent as a replacement for water. ASTM C188 classified the specific gravity of various and common mineral admixtures such as FA, PC, and GGBFS as follows: 2.0–2.7, 3.0–3.20, and 2.9–3.0, respectively [7]. Sabat [26] assumed that FA could be the most suitable geotechnical material, due to its resistivity in terms of high shear strength, low specific gravity, less compressibility, and good physicochemical properties. FA mainly contains silica, alumina, iron, and calcium, with less quantities of magnesium, sulfur, sodium, potassium, and carbon. The density or specific gravity of FA depends on its chemical compounds and typically ranges between 1.9 and 2.8 [27].

2.2.3 Size and fineness of particle

As mentioned before, FA particles have spherical solid forms with hollowing inside as cenospheres or plerospheres form. FA particle sizes vary from 1 μm to more than 100 μm. In general, 10–30% of particles are larger than 45 μm, with 300–500 m^2/kg of surface area. However, some types of FA have low or high surface area between 200 m^2/kg and 700 m^2/kg, respectively [27]. There are two ways to measure the particle size and fineness of FA:

- Specific surface area by Blaine apparatus: this method is based on the time passing through a bed of FA and correlated with its specific surface area in m^2/kg.

  ASTM does not exaggerate any specific requirement for the surface area of FA, which could be used in concrete, whereas the Indian Standard IS 3812 Part 1 [28] specifies 320 m^2/kg of FA as a minimum Blaine area for use in concrete.

- Residue on 45 μm sieve by wet-sieve analysis: this method is used to measure the percentage of particles in FA bigger than 45 μm as is referred to in ASTM430 [29]. Many countries follow this method for their national standards [7].

Some research showed that particles of raw FA mostly range from 1 to 100 μm in **Figure 2**. The particles less than 10 μm are the ones that react and contribute in the formation of early strength (7 and 28 days), whereas the particles between 10 and 45 μm react slowly and lead to the formation of a late strength (up to 1 year). The particles higher than 45 μm could be considered as inert and largely act as fine sand (filler) [7, 27].

2.2.4 Color

FA from bituminous coal has a darker gray color which comes from lignite or sub-bituminous coal and also can be buff to tan in color. It is thought that the gray color could be explained by the presence of unburned carbon (UBC). If the

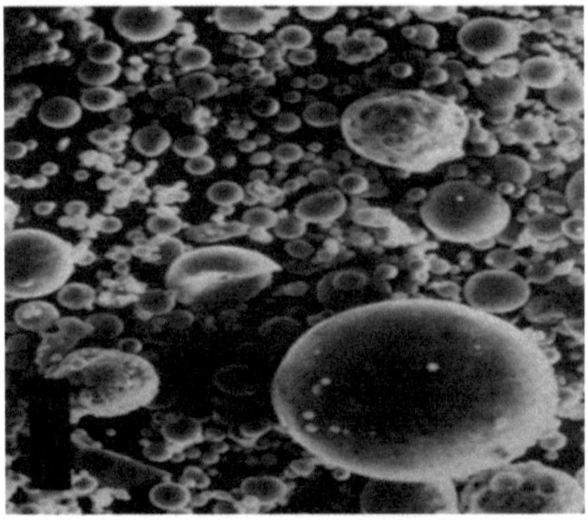

Figure 2.
Scanning electron microscope (SEM) micrographs of fly ash particles [30].

percentage of carbon is low or absent in ash, then the color might be brown, due to the presence of iron (+3) compounds. The color changes to bluish gray to gray if the iron compounds are (+2) [7].

Tanosaki et al. [13] have reported the use of colorimetric methods or as is known as the Munsell system to identify colors by following the next three dimensions, hue, value (lightness), and chroma (color purity). In 1905, Professor Albert H. Munsell created the Munsell system. According to Malacara [31], **Figure 3**, describes the color circle system. The system is divided into five principle hues: red, yellow, green, blue, and purple, along with five intermediate hues halfway between adjacent principle hues. Each of these ten categories is used to divide into other ten sub-categories, so that 100 hues are given integer values.

2.2.5 Unburned carbon

Unburned carbon (UBC) is mostly the significant affecting particles on the loss on ignition (LOI). During hydration process the carbon particles do not have any part in the chemical reactions. However, they have an impact on the water requirement in concrete. Carbon particles have a very strong affinity and attraction to

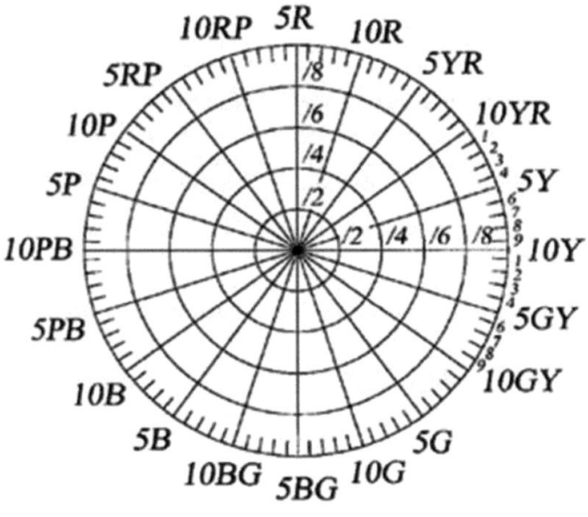

Figure 3.
Munsell color circle [31].

the organic chemical admixtures. For example, air-entraining agents (AEA) are a chemical agent that has absorbed on the carbon and negatively affects the hardened concrete. In general, the absorption degree depends on many factors, such as surface area and type of carbon in terms of its polarity and particle size. An experimental study showed that FA with less than 3–4% of carbon does not have a greater effect on the performance of organic chemical admixtures [7]. On the other hand, Ha et al. [32] reported that the use of FA as raw material, which contains around 8% of UBC, could accelerate the corrosion of reinforcement steel.

2.3 Chemical and mineralogical composition of FA

FA has varieties of chemical compositions; the averages of the main elements of FA in some European countries (France, UK, Germany), USA, and far Asian countries (Japan, China, India) are given as follows: 53.05%, 27.24, and 5.50% for SiO_2, Al_2O_3, and Fe_2O_3, respectively [7]. However, another study [33] reported that FA consisted of a heterogeneous mixture of complex aluminosilicate glasses and other crystalline elements. The structure of aluminosilicate glass is an amorphous form, but it could be modified due to the addition of alkaline and metal oxides such as Na_2O, K_2O, MgO, CaO, and FeO. A study carried out by Das and Yudhbir [34] showed that a strong correlation exists between the glass content and the ratio of potassium to aluminum oxides (K_2O/Al_2O_3). ASTM classification shows that the composition of glass in class F fly ash is different from that in class C. A high polymerized glass network is observed in class F FA, but the glass matrix depolymerizes when the CaO increases in comparison with Al_2O_3 content [7, 21, 35].

2.4 Setting time

The addition of FA or other raw material such as GGBFS generally delays the setting time of concrete. The initial and final setting time averages of class F and class C FA are 4:50, 4:40 and 6:45, and 6:15 (h:min), respectively. Setting time could be affected by different factors, for example, the amount of Portland cement, water demand, the reactivity of the pozzolan dosage or FA, and the temperature of concrete. Hot weather plays a positive effect on setting times and is considered as an advantage, by giving enough time for placing and finishing the handled work. On the other hand, if the weather is cold, setting time could be controlled by additives, which delay the finish operation. Some of accelerating admixtures and calcined shale or clay could be used to decrease setting time [27].

FA might have an influence on the rate of the hardening of cement [30, 36, 37] for the following reasons:

- FA is considered as cementitious and contains high calcium (class C FA).
- FA could contain sulfates which lead to a reaction with cement in the same way as when gypsum is added to Portland cement.
- The fly ash cement mortar might contain less water, and this has a significant effect on the rate of stiffening.
- The surface-active agent which could be added to modify the rheology (water reducers) of concrete could be absorbed by FA, and this leads to an influence on the stiffness of mortar.
- FA particles could act as nuclei for crystallization of cement hydration products.

2.5 Physical treatment

A study carried out by Barry [38] shows that the CSA (Canadian Standards Association) standard A23.5-M1982 on plant scale gets an advantage by improving the quality of using FA with high finer size of particles. The results showed an improvement in terms of reactivity and activity of FA, reducing water requirement and resulting in an enhanced ability to control alkali-aggregate reaction. It is observed that the particle size (or the particle surface area) and the size distribution have a significant role in determining the activity of FA. Therefore, FA with finer particle size could replace a high proportion of cement without affecting the strength [7].

However, Ramezanianpour [30] and Adam [39] have performed more than 340 tests of 14 sources of FA. Their results showed that there is no correlation between fineness and compressive strength at the ages of 7 and 28 days for mortars, but a minor correlation was found at 90 days. Joshi [40] and Ravina [41] have exploited a new phenomenon which is called "particle size segregation phenomenon" electrostatic precipitators' method was used to obtain FA fractions of different fineness from a particular source. Another experiment was carried out by Joshi [40] by investigating the proportion of particles of four types of FA, which are up to 45 μm from a modern power plant. The retained percentages of 45 μm sieve for each type of FA are 5, 16, 32, and 38%. The results indicated that replacing 10 and 20% of finer FA in concrete leads to develop a significant strength. These results have been supported by those found by Ravina [41] when the pozzolanic activity index of low-calcium FA from the same precipitator was used for testing.

2.6 Effect of FA on workability and water requirement

In general, rheological properties of cement pastes could be impacted by the morphology and the small size of the spherical particles of FA. The amount of calcium in FA particles (low calcium) has a significant influence on the rheology of pastes by reducing the amount of water demand and increasing workability. According to Davis et al. [42], FA is considered as a particular material comparing with other pozzolans by leading to the increased water requirement of concrete mixtures. Owens [43] believed that the main characteristic of FA, which has a significant effect on workability of concrete, is the proportion of coarse material (up to 45 μm) which could exist in FA. The effect of coarse particles on the water requirement is shown in **Figure 4**.

Much research has been carried out by Lloyd and Rangan [44] on the use of FA in geopolymer concrete. It is investigated that not only compressive strength could be affected by the characteristics of initial materials but also workability of geopolymer concrete. However, other studies [45] have shown that workability might be related to the ratio of alkaline activator solution (AAS) to binder and composition and nature of the chemical admixture which has been used.

Sathia et al. [46] have reported that the ratio of H_2O to Na_2O of 10–14 is only used when FA content is about 408 kg/m^3 in a designed concrete; this ratio could be changed depending on FA content. Thus, Siddique and Iqbal Khan [47] stated that for an equal w/c ratio and depending on the spherical shape and glassy phase on the FA particle surface, a greater workability could be achieved. Ramezanianpour [30] stated that, due to the necessity of mixing and placing concrete in a reinforced formwork, it is necessary to maintain its workability. This could be determined by the rheological properties of the system, which are in turn impacted by all the components. Thus, it is important to understand the rheological behavior and the main role of FA in the fresh concrete, which leads to exploit the potential role of FA for improving concrete.

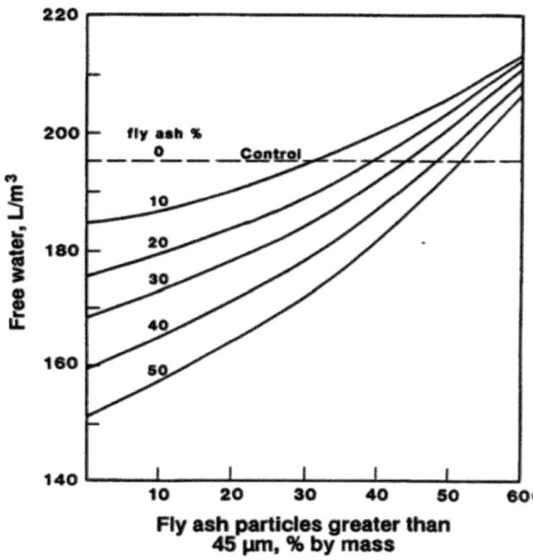

Figure 4.
Influence of coarse-particulate content of FA on water requirement for equal workability in concrete [43].

2.7 The impact of FA on durability of concrete exposed to elevated temperatures

Recently, the requirement of infrastructure and its development such as in nuclear reactor containment structures exaggerates the use of concrete, which could withstand high temperatures. Many researchers [48–50] have studied the effect of elevated temperature on FA concrete in the range of 230°C. Another study was carried out by Carette et al. [51] which showed the influence of a temperature 600°C on concrete with a mix of Portland cement, slag, and FA, as is illustrated in **Figure 5**. Under a high temperature, the addition of FA has no effect on the behavior of the concrete; however the changes in concrete properties or decreasing of strength could be observed at the same range of temperatures [37].

In addition, degradation of concrete structures is strongly affected by the chemical attack. For example, the penetration of chloride ions into the concrete leads to chemical reactions which could help in the formation of corrosion around reinforcement. This could be the reason of an early end to a structure's life cycle. Other studies that have been carried out by Thomas et al. [52] and Uddin and Shaikh [53] have reported that resistance of concrete to the immigration of chloride ions is mainly controlled by porosity and inter-connectivity of pores system and also depends to the chemical binding capacity of cement.

2.8 FA requirements for geopolymer

In order to achieve an efficient geopolymer synthesis, it is required that silica (SiO_2), alumina (Al_2O_3), and iron (Fe_2O_3) should be in high proportions [54]. Further, the activity of FA or the formation of aluminosilicate gel is related to the nature of environment, which could be acidic or basic (Ferna and Deventer, 2007) [55], and also high concentration of calcium has an important effect on the reaction, by accelerating its rate. Nikolić et al. [56] reported that the reactivity of FA could be influenced by many factors which in turn affects the characteristics of FA-based geopolymer, such as glassy phase, particle size distribution, the presence of iron, calcium, and inert elements. However, the reactivity of FA is not dependent only on the glassy phase but on the whole FA; this means that the glassy phase has a limitation

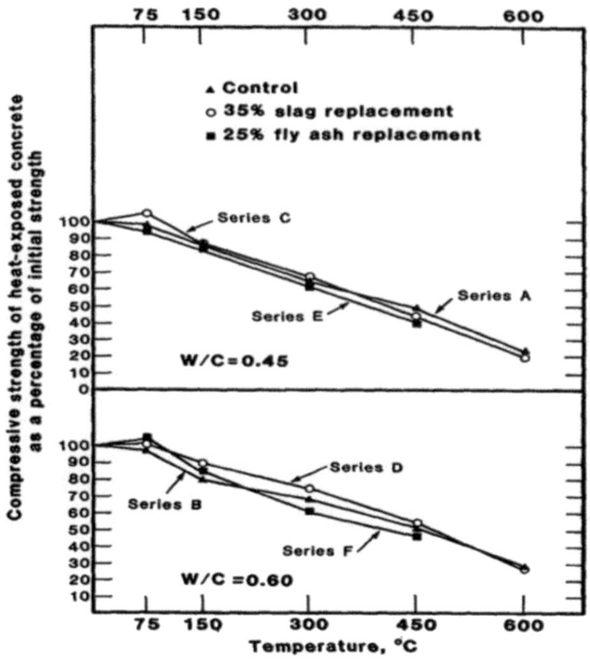

Figure 5.
Compressive strength of concretes after 1 month of exposure to various elevated temperatures [51].

degree. Therefore, the reactivity of FA usually depends on the dissolution level of FA in the alkaline activator. As aforementioned, loss on ignition (LOI) is defined as the unburned carbon present in FA and how that affects the quality of paste or concrete by increasing the water requirement and reducing the reactivity of pozzolanic constituents. ASTM C618 (2008) [39] has reported that the required percentage of LOI is limited to 6% maximum. Another study showed that a high proportion of SO_3 in concrete could lead to instability in volume, which in turn has an impact on durability. However, it is reported that about 5% of SO_3 of FA could be used as a concrete binder.

2.9 Chemical activation

Blanco et al. [57] have proposed a procedure of using of wet milling and leaching with sulfuric acid to activate FA. One of the main applications to use the activated FA is to substitute silica fume in concrete, which could lead to achieve a high strength, due to the decrease of pore size in the hardened concrete. The addition of a limited amount of sodium sulfate or potassium sulfate (Na_2SO_4 or K_2SO_4) mixed with calcium hydroxide ($Ca(OH)_2$) has a substantial effect on acceleration of hydration and compressive strength. A study carried out by Görhan and Kürklü [58] investigated that the activation of mortar samples by using NaOH of 6 M leads to increase in compressive strength values by 21.3 MPa and 22 MPa, compared to those samples which are activated by 9 M NaOH. Therefore, it has been reported by other authors [59, 60] that to achieve a great reactivity of FA particles within the activator solution, the liquid phase plays a significant role as a transport medium and a less smoothly gel is formed, due to the faster reaction of NaOH.

2.10 Addition of FA to cement and concrete

FA is classified into two classes by ASTM C618. Class F FA is pozzolanic, with minimum or no cementing value, whereas class C FA is cementitious as the same as

Characteristics	ASTM C618 FA
Type	ASTM class F and C
Shape and particles	Cenospheres, plerospheres, spherical particles
Specific gravity	2.0–2.7
Fineness	1–100 μm
Color	Darker gray buff to tan
LOI	Depends to the source
Mineral compositions	Rich in aluminosilicate, iron oxide, and calcium oxide
Thermal resistance	High resistance to elevated temperatures
Uses in concrete production	Improves concrete's strength and durability

Table 3.
FA characteristics.

pozzolanic properties [6]. The main parameter to formulate a concrete mix design with the addition of FA is the proportion of the mix under consideration of the variation of water-cementitious ratio. This could lead to achieve the requirements for compressive strength at different ages, air content, and workability. ACI 211.1 or 211.2 has determined the procedures for the mix design in details, in terms of the proportioning of water, cement (or cement plus FA), and aggregate materials [61].

However, the specific gravity of FA is lower than PC, which needs to be taken into consideration in the mix proportioning process. Other standards such as the European standard BS EN206 provide some requirements for the use FA in concrete [7, 27, 62]. Further study has been carried out by Horpibulsuk et al. [63] which reported that FA could be considered as a dispersing material, when it is mixed with cement. This in contrast when FA is used in concrete as a pozzolanic material, where a pozzolanic reaction can be occurred by consuming of $Ca(OH)_2$ during the hydration process. **Table 3** summarizes the main physical and chemical characteristics of FA that can be referred to in any uses.

3. Conclusion

This chapter mainly reviews FA as basic raw materials, which can be introduced in the production of concretes so-called geopolymer concrete. As per several reports, a huge quantity of FA is disposed and landfilled. Besides, it has been estimated that the cement industry contributes approximately 7% of global warming, due to the substantial increase in carbon dioxide. The raw materials and the manufacturing process of the conventional cement are found to be the main reason behind this increase. As a solution for this serious environmental issue, much research has been conducted to investigate other alternative materials to Portland cement. FA is the most investigated material, due to its suitability and its physicochemical properties including microstructure, reaction mechanism, and characterization as normalized by ASTM C618.

The chemical composition analysis shows that FA consists of a complex oxide mixture of aluminosilicate glasses and other crystalline elements with the presence of the amorphous phase. This latter explains the high reactivity of FA and its suitability as a mineral admixture for cement and concrete.

Numerous studies show that including FA in the production of geopolymers concrete provides a greater mechanical and microstructure properties, due to its physical characteristics compared to that given by OPC. These findings make the use of FA of particular interest to researchers. As a result of the excellent properties, geopolymer-based FA has been successfully applied in various traditional and new applications. It is believed that the progress researches in geopolymers area especially in the utilization of industrial by-products have been intensified and consisted as a significant step to introduce geopolymers technology in the construction industry and particularly in marine applications.

Acknowledgements

The authors would like to acknowledge the funding support from "Partnership for Research in Geopolymer Concrete" (H2020-MSCA-RISE-2015-689857-PRIGeoC) sponsored by the European Union and the Centre of Excellence Geopolymer & Green Technology (CeGeoGTech), Universiti Malaysia Perlis (UniMAP), in providing us this research opportunity and supporting us with all the necessary facilities.

Conflict of interest

The authors declare no conflict of interest.

Nomenclature

FA	fly ash
PFA	pulverized fly ash
ASTM	American Society for Testing and Materials
GGBFS	ground-granulated blast furnace slag
OPC	ordinary Portland cement
PC	Portland cement
CSA	Canadian Standards Association
AEA	air-entraining agents
UBC	unburned carbon
LOI	loss on ignition

Author details

Aissa Bouaissi[1*], Long Yuan Li[1], Mohd Mustafa Al Bakri Abdullah[2], Romisuhani Ahmad[2], Rafiza Abdul Razak[2] and Zarina Yahya[2]

1 School of Marine Science and Engineering, University of Plymouth, United Kingdom

2 Center of Excellence Geopolymer and Green Technology (CEGeoGTech), School of Materials Engineering, Universiti Malaysia Perlis (UniMAP), Kangar, Perlis, Malaysia

*Address all correspondence to: aissabouaissi@yahoo.fr

References

[1] Neupane K. Fly ash and GGBFS based powder-activated geopolymer binders: A viable sustainable alternative of portland cement in concrete industry. Mechanics of Materials. 2016;**103**:110-122. DOI: 10.1016/j.mechmat.2016.09.012

[2] Abdullah MMA, Kamarudin H, Nizar IK, Sandu AV, Binhussain M, Zarina Y, et al. Design, processing and characterization of fly ash-based geopolymers for lightweight concrete application. Revista de Chimie (Bucharest). 2013;**64**(4):382-387

[3] Davidovits J. Geopolymers: Inorganic polymerie new materials. Journal of Thermal Analysis. 1991;**37**:1633-1656

[4] Pavithra P, Srinivasula Reddy M, Dinakar P, Hanumantha Rao B, Satpathy BK, Mohanty AN. A mix design procedure for geopolymer concrete with fly ash. Journal of Cleaner Production. 2016;**133**:117-125. DOI: 10.1016/j.jclepro.2016.05.041

[5] Hamidi RM, Man Z, Azizli KA. Concentration of NaOH and the effect on the properties of fly ash based geopolymer. Procedia Engineering. 2016;**148**:189-193. DOI: 10.1016/j.proeng.2016.06.568

[6] ASTM. C618-08a: Specification for coal Fly ash and raw or Calcined natural Pozzolan for use in concrete. ASTM International. 2008. DOI: 10.1520/c0618-03

[7] Jayant D. Mineral Admixtures in Cement and Concrete. Boca Raton, New York: Taylor & Francis Group, LLC, CRC Press; 2013

[8] Fernández-Jiménez A, Palomo A, Criado M. Microstructure development of alkali-activated fly ash cement: A descriptive model. Cement and Concrete Research. 2005;**35**(6):1204-1209

[9] Hajimohammadi A, Provis JL, van Deventer JSJ. The effect of silica availability on the mechanism of geopolymerisation. Cement and Concrete Research. 2011;**41**(3):210-216. DOI: 10.1016/j.cemconres.2011.02.001

[10] Diaz EI, Allouche EN, Eklund S. Factors affecting the suitability of fly ash as source material for geopolymers. Fuel. 2010;**89**(5):992-996. DOI: 10.1016/j.fuel.2009.09.012

[11] Salloum T. Effect of Fly Ash Replacement on Alkali and Sulphate Resistance of Mortars. 2007. Available from: https://spectrum.library.concordia.ca/975471/

[12] Brahammaji GSV, Muthyalu PV. A study on performance of fly ash based geopolymer concrete in chemical atmosphere. International Journal of Advanced Engineering and Technology. 2015;**8**(4):574-586

[13] Tanosaki T, Yu Q, Nagataki S. Are fly shes green? In: Proceedings of 2011 World of Coal Ash (WOCA) Conference, Denver, CO, USA; 9-12 May 2011. pp. 1-5

[14] Yahya Z, Abdullah MMA, Kamarudin H, Khairul NI, Sandu AV, Petrica V, et al. Chemical and physical characterization of boiler ash from palm oil industry waste for geopolymer composite. Revista de Chimie. 2013;**64**(12):32-34

[15] Van Jaarsveld JGS, Van Deventer JSJ, Lukey GC. The characterisation of source materials in fly ash-based geopolymers. Materials Letters. 2003;**57**:1272-1280

[16] Komljenović M, Baščarević Z, Bradić V. Mechanical and microstructural properties of alkali-activated fly ash geopolymers. Journal of Hazardous Materials. 2010;**181**(1-3):35-42. DOI: 10.1016/j.jhazmat.2010.04.064

[17] Van Riessen A, Chen-Tan N. Beneficiation of Collie fly ash for synthesis of geopolymer: Part 1—Beneficiation. Fuel. 2013;**106**:569-575. DOI: 10.1016/j.fuel.2012.11.070

[18] Marjanović N, Komljenović M, Baščarević Z, Nikolić V. Improving reactivity of fly ash and properties of ensuing geopolymers through mechanical activation. Construction and Building Materials. 2014;**57**:151-162. DOI: 10.1016/j.conbuildmat.2014.01.095

[19] Nikolić V, Komljenović M, Marjanović N, Baščarević Z, Petrović R. Lead immobilization by geopolymers based on mechanically activated fly ash. Ceramics International. 2014;**40**(6):8479-8488. DOI: 10.1016/j.ceramint.2014.01.059

[20] Komljenovic M, Bašc Z, Marjanovic N. The influence of fly ash characteristics and reaction conditions on strength and structure of geopolymers. Construction and Building Materials. 2015;**94**:361-370

[21] Brouwers HJH, Van Eijk RJ. Chemical Reaction of Fly Ash. 2003. pp. 11-16. Available from: http://doc.utwente.nl/45282/1/Conference12.pdf

[22] Atiş CD, Sevim UK, Özcan F, Bilim C, Karahan O, Tanrikulu AH, et al. Strength properties of roller compacted concrete containing a non-standard high calcium fly ash. Materials Letters. 2004;**58**(9):1446-1450. DOI: 10.1016/j.matlet.2003.10.007

[23] Givi AN, Rashid SA, Aziz FNA, Salleh MAM. Assessment of the effects of rice husk ash particle size on strength, water permeability and workability of binary blended concrete. Construction and Building Materials. 2010;**24**(11):2145-2150

[24] Feng X, Clark B. Evaluation of the physical and chemical properties of fly ash products for use in Portland cement concrete. In: 2011 World of Coal Ash Conference, Denver, CO, USA; 9 May 2011

[25] Ferdous MW, Kayali O, Khennane A. A detailed procedure of mix design for fly ash based geopolymer concrete. In: Fourth Asia-Pacific Conf FRP Struct (APFIS 2013). 2013. pp. 11-13

[26] Sabat AK. Prediction of maximum dry density and specific gravity of Fly ash using support vector machine. EJGE. 2015;**20**:155-166

[27] Malvar LJ, Cline GD, Burke DF, Rollings R, Greene J, Sherman TW. Alkali-Silica Reaction Mitigation: State-of-the-Art. Defense Technical Information Center; 2001. DOI: 10.21236/ada526563

[28] IS 3812 (Part 1): 2003 Indian Standard Pulverized Fuel Ash—Specification (Second Revision). 2003. p. 3812

[29] ASTMC430. Standard Test Method for Fineness of Hydraulic Cement by the 45-Mm (No. 325) Sieve. ed West Conshohocken: PA ASTM Int; 2003

[30] Ramezanianpour AA. Cement Replacement Materials. 2014. Available from: http://link.springer.com/10.1007/978-3-642-36721-2

[31] Malacara D. Color Vision and Colorimetry: Theory and Applications. 2nd ed. Bellingham, WA: SPIE; 2011. DOI: 10.1117/3.881172

[32] Ha T-H, Muralidharan S, Bae J-H, Ha Y-C, Lee H-G, Park KW, et al. Effect of unburnt carbon on the corrosion performance of fly ash cement mortar. Construction and Building Materials. 2005;**19**(7):509-515. DOI: 10.1016/j.conbuildmat.2005.01.005

[33] Bapat JD. Performance of cement concrete with mineral admixtures.

Advances in Cement Research. 2001;**13**(4):139-155. DOI: 10.1680/adcr.13.4.139.39286

[34] Das SK, Yudhbir. A simplified model for prediction of pozzolanic characteristics of fly ash, based on chemical composition. Cement and Concrete Research. 2006;**36**(10):1827-1832. DOI: 10.1016/j.cemconres.2006.02.020

[35] Malhotra VM. Fly Ash in Concrete. Energy, Mines and Resources Canada. CANMET= Énergie, Mines et Ressources. 1986

[36] Chindaprasirt P, Jaturapitakkul C, Sinsiri T. Effect of fly ash fineness on microstructure of blended cement paste. Construction and Building Materials. 2007;**21**(7):1534-1541

[37] Berry EE, Hemmings RT, Zhang MH, Malhotra VM, Bilodeau A, Carette GG. Investigation of high-volume fly ash concrete systems. In: Final Report. 1993

[38] Barry EE. Beneficiated fly ash: Hydration, microstructure, and strength development in Portland cement systems. Special Publication. 1989;**114**:241-274

[39] Adam AA. Strength and durability properties of alkali activated slag and fly ash-based geopolymer concrete [PhD thesis]. RMIT Univ; 2009

[40] Joshi RC. Effect of coarse fraction of fly ash on concrete properties. In: Proceedings of 6th International Symposium on Fly Ash Utilization. 1982. pp. 52-82

[41] Ravina D. Production and collection of fly ash for use in concrete. In: Proc Symp Fly Ash Inc Hydrated Cem Syst Bost. 1981. pp. 2-11

[42] Davis RE, Carlson RW, Kelly JW, Davis HE. Properties of cements and concretes containing fly ash. ACI Journal Proceedings. 1 May 1937;**33**(5):577-612

[43] Owens PL. Fly ash and its usage in concrete. Journal of Concrete Society. 1979;**13**(7):21-26

[44] Lloyd N, Rangan BV. Geopolymer concrete: A review of development and opportunities. CipremierCom. 2010

[45] Sabitha D, Dattatreya JK, Sakthivel N, Bhuvaneshwari M, Sathik SJ. Reactivity, workability and strength of potassium versus sodium-activated high volume fly ash-based geopolymers. Current Science. 10 Dec 2012. pp. 1320-1327

[46] Sathia R, Babu Ganesh K, Santhanam M. Durability study of low calcium fly ash. In: 3rd ACF Int Conf 2008 D21. 2008

[47] Siddique R, Iqbal Khan M. Fly Ash Supplementary Cementing Materials. Berlin Heidelberg: Springer; 2011. pp. 1-66. DOI: 10.1007/978-3-642-17866-5_1

[48] Nasser KW, Marzouk HM. Properties of concrete made with sulfate resisting cement and fly ash. Special Publication. 1983;**79**:383-396

[49] Nasser KW, Lohtia RP. Creep of mass concrete at high temperature. ACI Journal of Proceedings. 1971;**68**(4). DOI: 10.14359/11329

[50] Nasser KW, Marzouk HM. Properties of mass concrete containing fly ash at high temperatures. ACI Journal of Proceedings. 1 Apr 1979;**76**(4):537-550

[51] Carette GG, Painter KE, Malhotra VM. Sustained high temperature effect on concretes made with normal Portland cement, normal Portland cement and slag, or normal Portland cement and fly ash. Concrete International. 1982;**4**(7):41-51

[52] Thomas C, Setién J, Polanco JA, Alaejos P, Sánchez De Juan M. Durability of recycled aggregate concrete. Construction and Building Materials. 2013;**40**:1054-1065. DOI: 10.1016/j.conbuildmat.2012.11.106

[53] Uddin F, Shaikh A. Mechanical and durability properties of fly ash geopolymer concrete containing recycled coarse aggregates. International Journal of Sustainable Built Environment. 2016;**5**(2):277-287. DOI: 10.1016/j.ijsbe.2016.05.009

[54] Part WK, Ramli M, Cheah CB. An overview on the influence of various factors on the properties of geopolymer concrete derived from industrial by-products. Construction and Building Materials. 2015;**77**:370-395. DOI: 10.1016/j.conbuildmat.2014.12.065

[55] Duxson P, Fernández-Jiménez A, Provis JL, Lukey GC, Palomo A, van Deventer JSJ. Geopolymer technology: The current state of the art. Journal of Materials Science. 2006;**42**(9):2917-2933. DOI: 10.1007/s10853-006-0637-z

[56] Nikolić V, Komljenović M, Baščarević Z, Marjanović N, Miladinović Z, Petrović R. The influence of fly ash characteristics and reaction conditions on strength and structure of geopolymers. Construction and Building Materials. 2015;**94**:361-370. DOI: 10.1016/j.conbuildmat.2015.07.014

[57] Blanco F, Garcia MP, Ayala J, Mayoral G, Garcia MA. The effect of mechanically and chemically activated fly ashes on mortar properties. Fuel. 2006;**85**(14-15):2018-2026. DOI: 10.1016/j.fuel.2006.03.031

[58] Görhan G, Kürklü G. The influence of the NaOH solution on the properties of the fly ash-based geopolymer mortar cured at different temperatures. Composites Part B: Engineering. 2014;**58**:371-377. DOI: 10.1016/j.compositesb.2013.10.082

[59] Duchesne J, Duong L, Bostrom T, Frost R. Microstructure study of early in situ reaction of fly ash geopolymer observed by environmental scanning electron microscopy (ESEM). Waste and Biomass Valorization. 2010;**1**(3):367-377

[60] Abdullah MMA, Hussin K, Bnhussain M, Ismail KN, Ibrahim MW. Mechanism and chemical reaction of fly ash geopolymer cement—A review. International Journal of Pure and Applied Sciences and Technology. 2011;**6**(1):35-44

[61] Dixon DE, Prestrera JR, Burg GRU, Chairman SA, Abdun-Nur EA, Barton SG, et al. Standard practice for selecting proportions for Normal, heavyweight, and Mass Concrete (ACI 211.1-91). 1991

[62] Bouaissi A, Li LY, Moga LM, Sandu IG, Abdullah MMA, Sandu AV. A review on fly ash as a raw cementitious material for geopolymer concrete. Revista de Chimie. 2018;**69**(7):1661-1667

[63] Horpibulsuk S, Rachan R, Raksachon Y. Role of fly ash on strength and microstructure development in blended cement stabilized silty clay. Soils and Foundations. 2009;**49**(1):85-98

Chapter 8

Improvement of the Thermal Properties of Sorel Cements

Rim Zgueb, Amal Brichni and Noureddine Yacoubi

Abstract

Sorel cements is a promising building material for insulation applications. Indeed, the effect of polyvinyl acetate polymer on cements has been investigated. The polyvinyl acetate polymer was added to the cement matrix with a percentage of 0, 5, 10, 15 and 20% by weight of Sorel cement. The thermal properties of Sorel cement were determined by photothermal deflection technique. Thermal properties such as thermal conductivity and thermal diffusivity are measured by coincidentally the experimental curves of the photothermal signal with the best corresponding theoretical curves. The results revealed that the incorporation of polyvinyl acetate polymer enhance the thermal insulation and reduce the compressive strength of Sorel cement.

Keywords: Sorel cement, photothermal deflection technique, thermal insulation, mechanical properties, polyvinyl acetate polymer

1. Introduction

Thermal insulation consists in using a material or a combination of materials in order to limit heat losses by conduction, convection and radiation between the interior and exterior of a building due to its low thermal conductivity. Thus, thermal insulation acts as a thermal barrier. The two main criteria for thermal insulation of buildings are the optimization of energy consumption and the protection of buildings against climatic factors. In some countries, people spend up to 90% of their time indoors (offices, factory, workshops, homes, ...) so they need a viable artificial environment that requires energy. Thermal insulation reduces heat loss, saves heating, limits greenhouse gas emissions, and improves living comfort.

In addition, cement will remain the main material that meets the needs of modern infrastructure and participates in the construction of large structures. Thus the development of cement performance to meet current needs has prompted several research projects to improve its thermal insulation without reducing its resistance.

One of the most promising cements for thermal insulation is magnesium oxychloride cement (MOC). During the last decade, the development of magnesium oxychloride cement was motivated by environmental considerations. The MgO production temperature was lower than the conversion temperature of $CaCO_3$ into Portland cement. The energy savings associated with this reduced temperature have led many researchers to believe that magnesium oxychloride cements are the future of green cement. In addition, MgO has the capacity to potentially absorb CO_2

unlike Portland cement. It is “carbon neuter” cement. These two interdependent aspects have a recent academic and commercial interest in the field of MgO cements.

MOC cement is synthesized by dissolving the magnesia MgO into aqueous solution of magnesium chloride hexahydrate $MgCl_2$, forming a homogeneous gel from which the basic salts of magnesium chloride precipitate. These salts are expressed by $xMg(OH)_2.yMgCl_2.zH_2O$ phases which depend on the temperature, the reactivity of magnesium and the relationships between the number of moles of $MgCl_2$ to that of the water molecules, the temperature and the reactivity of magnesium [1]. It has several performances and has become popular due to its attractive appearance, similar to that of marble. Indeed, it rapid hardening rate and it has high mechanical strength and good resistance to abrasion. The abrasion resistance is three times that of ordinary Portland cement. Generally, the compressive strength is greater than 50 MPa after curing for 28 days [2–7]. In addition, a peculiarity of this cement is its resistance to salt and saline solutions. The main applications used are architectural applications such as the construction of thermal and acoustical insulating panels [8], the construction of industrial floors and other prefabricated building boards [9].

In this chapter, we investigated the thermal properties of Sorel cement. The aim of this research was to improve the thermal insulation of Sorel cement by using a polyvinyl acetate (PVAc) polymer. We have determined the thermal conductivity and thermal diffusivity of various blended PVAc cements and we are studying the effect of PVAc on the compressive strength of composite materials.

2. Experimental procedures

2.1 Specimen preparation

The main materials used to synthesize the MOC are Magnesia (MgO), magnesium chloride ($MgCl_2$) and water (H_2O). Magnesium oxide powder used in this study is mostly produced by calcinations of magnesite powder (HiMedia, India) at a temperature around 900°C. We dissolved magnesium chloride hexahydrate (Scharlab, Spain) in distilled water to prepare a saturated solution of magnesium chloride. The mass concentration of the solution was 217 g for 100 g of water. The mass ratio of $MgCl_2.6H_2O/MgO = 2.22$ [10]. The PVAc polymer (SICOP, Tunisia) is a fluid of low cost whose viscosity at 20°C is 10,000 $\pm$ 500 Pa s and the PH = 6 $\pm$ 1.

The samples were prepared by mixing at the same time magnesium oxide powder, saturated magnesium chloride solution and PVAc polymer. The polyvinyl acetate polymer is replaced by MgO. Five samples are synthesized; four samples with the incorporation of PVAc and a control sample (without addition). The obtained samples which are SC0 (0% PVAC), SC5 (5% PVAC), SC10 (10% PVAC), SC15 (15% PVAC) and SC20 (20% PVAC). The mixtures were cast into the cylindrical molds of height 50 mm and diameter about 25 mm, stored for 24 h, then unmolded and air-cured for 28 days.

2.2 Methods

The compressive strength of specimens was tested using Liyold mechanical testing instrument with a load of 300 KN. At the age of 28 days, the compressive strength for samples with diameter 25 mm and height 50 mm were measured at a loading speed of 2 mm/min at ambient temperature.

Measurement of thermal conductivity was performed in dry state using the photothermal deflection technique.

The micro-morphology on the fractured surface of MOC was characterized by scanning electron microscopy (SEM, JEOL-JSM- 5400). The sample is previously coated with a layer of gold and SEM was operated at 15 kV of acceleration voltage.

3. Photothermal deflection technique

3.1 Principle of the photothermal deflection technique PTD

The sample is heated by modulated and uniform light pump beam. The optical absorption of the sample will generate a unidimentional thermal wave that will propagate into the sample and in the surrounding fluid near the surface of the sample, inducing a temperature gradient then a refractive index gradient in the fluid. The absorbed light is transformed into heat by a nonradiative de-excitation process. A laser beam skimming parallely the sample surface and passing through this refractive index gradient is deflected. This deflection is related to the thermal and optical properties of the sample.

The deflection of the probe laser beam ψ is complex number $(\psi = |\psi|\, e^{j\varphi})$ given by [11]:

$$\psi(z,\ t) = \frac{L}{n_0}\frac{dn}{dT_f}\frac{\sqrt{2}}{\mu_f}\ |T_0|\ e^{-\frac{z_0}{\mu_f}}\ e^{j\left(\theta + \frac{\pi}{4} - \frac{z_0}{\mu_f}\right)}\ e^{j\omega t} \tag{1}$$

T_0 which is the periodic temperature rise at the sample surface is a complex number that written $T_0 = |\,T_0\,|\,e^{j\theta}$, Z_0 is the distance between the probe laser beam axis and the sample surface, L is the sample length in the direction of the laser probe beam, n is the fluid refractive index. Where $(\mu_f = D_f/\pi f)^{1/2}$ is the thermal diffusion length of the fluid with D_f the thermal diffusivity of the fluid and $j^2 = -1$.$|\psi|$ and φ are respectively the amplitude and the argument of the laser pump beam deflection given by:

$$|\psi(z)| = -\frac{L}{n_0}\frac{dn}{dT_f}\frac{\sqrt{2}}{\mu_f}\ |T_0|\ e^{-\frac{z_0}{\mu_f}} \tag{2}$$

$$\varphi = \frac{-z_0}{\mu_f} + \theta + \frac{\pi}{4} \tag{3}$$

In order to determine the deflection of the probe laser beam, we have to calculate the periodic temperature T_0 at the sample surface.

3.2 Calculation of the periodic elevation temperature T_0 at the sample surface

3.2.1 Bulk sample

3.2.1.1 Calculation of T_0

The sample consists of only one layer (**Figure 1**) of thermal conductivity k_s, thermal diffusivity D_s and thickness l_s. It is fixed at backing of k_b, D_b, l_b respectively thermal conductivity, thermal diffusivity and thickness. The sample and backing are in a fluid (air) of thermal conductivity k_f, thermal diffusivity D_f and thickness l_f. T_o determine

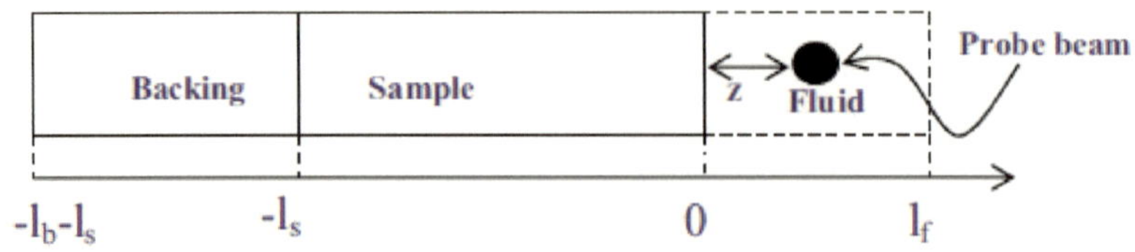

Figure 1.
Different medium browsed by the heat.

the periodic temperature T_0, we solve the one-dimensional heat diffusion equation in the different media backing, sample and fluid. Assuming that only the sample is absorbing the pump light beam, the heat equations in the different media are given by:

$$\begin{aligned} &\frac{\partial^2 T_f}{\partial z^2} = \frac{1}{D_f}\frac{\partial T_f}{\partial t} && \text{if } 0 \le z \le l_f \\ &\frac{\partial^2 T_s}{\partial z^2} = \frac{1}{D_s}\frac{\partial T_s}{\partial t} - Ae^{\alpha z}\left(1 + e^{j\omega t}\right) && \text{if } -l_s \le z \le 0 \\ &\frac{\partial^2 T_b}{\partial z^2} = \frac{1}{D_b}\frac{\partial T_b}{\partial t} && \text{if } -l_s - l_b \le z \le -l_s \end{aligned} \quad (4)$$

where $A = \alpha_s I_0/2ks$ are constant numbers and α_s is the optical absorption coefficient of the sample.

Indeed, the application of the boundary conditions of temperature and heat flux at the different interfaces allows us to determine the expression of the periodic temperature T_0 at the sample surface:

$$\begin{aligned} T_0 = -E\,&\left[(1-r)(1+b)\,e^{\sigma_s l_s} - (1+r)(1-b)\,e^{-\sigma_s l_s}\right. \\ &\left. + \,2(r-b)\,e^{-\alpha l_s}\right]/\left[(1+g)\,(1+b)\;e^{\sigma_s l_s} - (1-g)(1-b)\;e^{-\sigma_s l_s}\right] \end{aligned} \quad (5)$$

3.2.1.2 Illustration of theoretical model

We will study in this paragraph the variations of the photothermal signal as a function of the square root of the frequency in the case of a bulk sample. **Figure 2** represents the theoretical variations normalized amplitude and phase of the photothermal signal as a function of the square root of the frequency for three values of the thermal conductivity (k = 0.1, 2.0 and 4.0 W/m K) of a bulk sample.

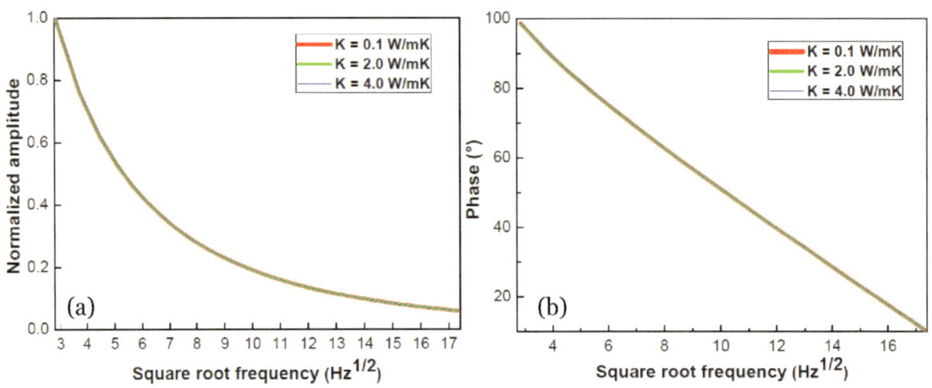

Figure 2.
Theoretical variation of the normalized amplitude (a) and phase (b) of the photothermal signal with the square root of modulation frequency for three numerical values of thermal conductivity.

We see that the normalized amplitude and the phase of the PTD signal are insensitive to variations of k (the three curves are coincident), which means that in this case we cannot determine the thermal conductivity of the sample.

Figure 3 represents the theoretical variations normalized amplitude and phase curves for different values of thermal diffusivity D. These curves show a remarkable sensitivity of the PTD signal to variations in thermal diffusivity, which allows determining thermal diffusivity with great precision.

3.2.2 Sample composed (layer deposed on a substrate)

3.2.2.1 Calculation of T_0

The sample is composed of a layer deposed on a substrate (**Figure 4**). Fernelius [10] was developed the theoretical model of a layer deposed on a substrate by writing the heat equations in the four medium: fluid (K_f, D_f, l_f), sample (K_s, D_s, l_s), black layer (K_c, D_c, l_c) and backing (K_b, D_b, l_b). Assuming that all the light is absorbed only by the black layer, the heat equations in the different media are given by:

$$\begin{aligned}
&\frac{\partial^2 T_f}{\partial z^2}=\frac{1}{D_f}\frac{\partial T_f}{\partial t} && if\ 0\le z\le l_f\\
&\frac{\partial^2 T_c}{\partial z^2}=\frac{1}{D_c}\frac{\partial T_c}{\partial t}-A_c e^{\alpha_c z}\left(1+e^{j\omega t}\right) && if\ -l_c\le z\le 0\\
&\frac{\partial^2 T_s}{\partial z^2}=\frac{1}{D_s}\frac{\partial T_s}{\partial t} && if\ -l_c-l_s\le z\le -l_s\\
&\frac{\partial^2 T_b}{\partial z^2}=\frac{1}{D_b}\frac{\partial T_b}{\partial t} && if\ -l_c-l_s-l_b\le z\le -l_c-l_s
\end{aligned} \tag{6}$$

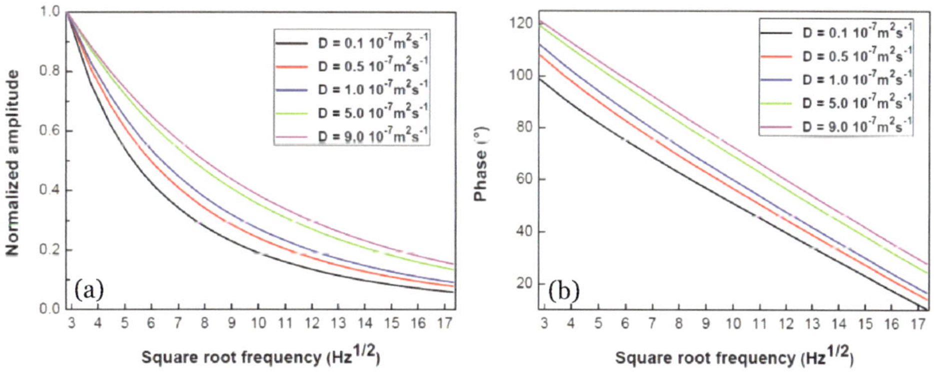

Figure 3.
Theoretical variation of the normalized amplitude (a) and phase (b) of the photothermal signal with the square root of modulation frequency for different values of thermal diffusivity.

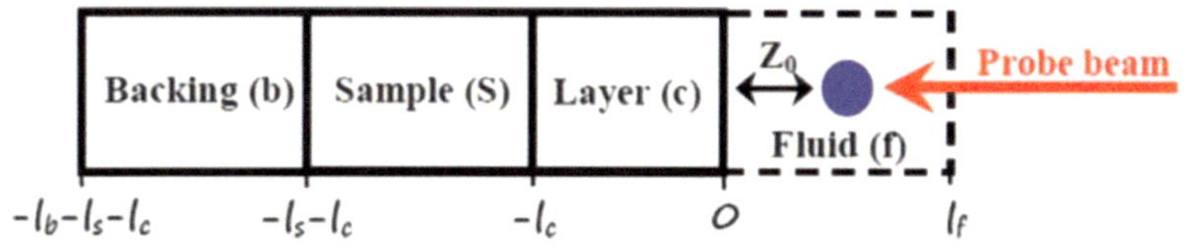

Figure 4.
Different medium.

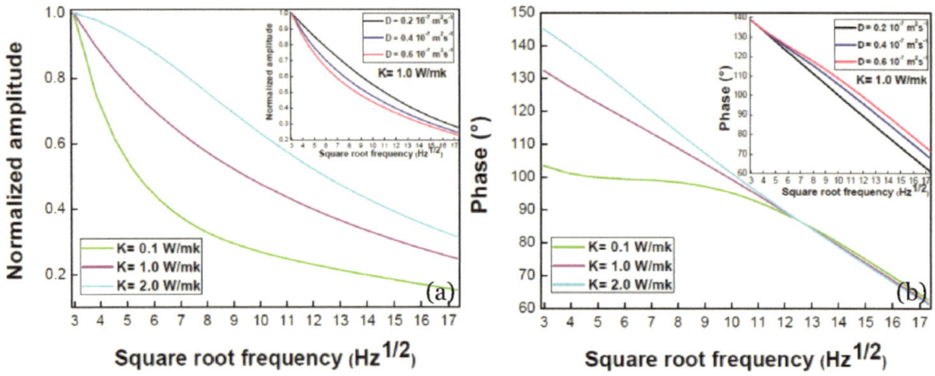

Figure 5.
Theoretical variation of the normalized amplitude (a) and phase (b) of the photothermal signal with the square root of modulation frequency for three values of thermal diffusivity and thermal conductivity.

where $A_c = \alpha_c I_0/2k_c$ are constant numbers and α_c is the optical absorption coefficient of the black layer.

By applying the continuity conditions of temperature and heat flow at the different interfaces allows us to determine the expression of the periodic temperature T_0 at the sample surface:

$$\begin{aligned}T_0 = {} & E_c[(1-b)e^{-\sigma_s l_s}[(1-r_c)(1-c)e^{\sigma_c l_c}+(1+r_c)(1+c)e^{-\sigma_c l_c} \\ & -2(1+c\,r_c)e^{-\alpha_c l_c}]-(1+b)e^{\sigma_s l_s}[(1-r_c)(1+c)e^{\sigma_c l_c}+(1+r_c)(1-c)e^{-\sigma_c l_c} \\ & -2(1-c\,r_c)e^{-\alpha_c l_c}]]\,/[(1+b)e^{\sigma_s l_s}\left[\left(1+\frac{g}{c}\right)(1+c)e^{\sigma_c l_c}+\left(1-\frac{g}{c}\right)(1-c)e^{-\sigma_c l_c}\right] \\ & -(1-b)e^{-\sigma_s l_s}\left[\left(1+\frac{g}{c}\right)(1-c)e^{\sigma_c l_c}+\left(1-\frac{g}{c}\right)(1+c)e^{-\sigma_c l_c}\right]]\end{aligned} \tag{7}$$

where $E_c = A_c/(\alpha_c^2 - \sigma_c^2)$, $r_c = \alpha_c/\sigma_c$, $b = k_b\sigma_b/k_s\sigma_s$, $c = k_c\sigma_c/k_s\sigma_s$, and $g = k_f\sigma_f/k_s$.

3.2.2.2 Illustration of theoretical model

Figure 5 shows the variations of normalized amplitude and phase for three values of thermal conductivity with equal thermal diffusivity $D = 0.4\ 10^{-7}\ m^2\ s^{-1}$ and well-defined values of the properties of the ink layer [11]. In addition, **Figure 5** also shows the variations normalized amplitude and phase for different values of thermal diffusivity with $k = 1\ Wm^{-1}\ K^{-1}$.

We notice that the PTD signal is sensitive to both the conductivity and the thermal diffusivity of sample (substrate). Since the value of thermal diffusivity is determined in the previous case (sample without layer), the value of thermal conductivity can be determined with great precision.

4. Determination of the thermal diffusivity of the samples (without black layer)

In this section, we will study the thermal diffusivity of samples (MOC without and with *PVAc*) using the Photothermal Deflection technique PTD. Thermal diffusivity is measured by the coincidence between the experimental curves of photothermal signal to the corresponding theoretical ones.

4.1 Thermal diffusivity of Sorel cement without PVAc

Figure 6 presents the experimental curves of normalized amplitude (a) and phase (b) of the photothermal signal and the corresponding theoretical ones for three values of the thermal diffusivity 0.2, 0.4 and 5.0 m^2/s versus square root modulation frequency for the reference sample (MOC without PVAc). Furthermore, **Figure 6** also presents the experimental normalized amplitude and phase of the photothermal signal and the corresponding theoretical ones for three values of the thermal conductivity 0.01, 0.3 and 0.7 w/mk versus the square root modulation frequency for magnesium oxychloride cement.

As noted in the previous section, the PTD signal is insensitive to thermal conductivity. We can only determine the values of the thermal diffusivity of the magnesium oxychloride cement. The theoretical curve which coincides best with the experimental curve is obtained for $0.4\ 10^{-7}\ m^2\ s^{-1}$.

4.2 Thermal diffusivity of Sorel cement with PVAc

We will proceed the same way for the rest of the magnesium oxychloride cement samples to which PVAc has been added with different percentages using the same method. **Figure 7** shows the normalized amplitude and phase of experimental

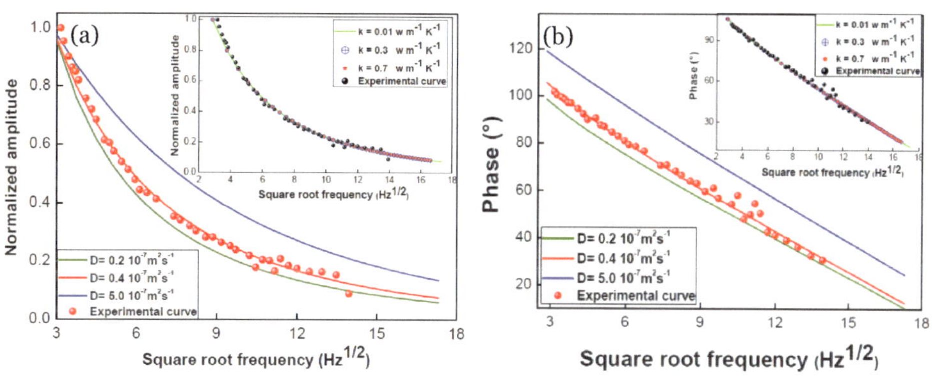

Figure 6.
Experimental and theoretical variation of the normalized amplitude (a) and phase (b) of the photothermal signal with the square root of modulation frequency for three values of thermal diffusivity and thermal conductivity.

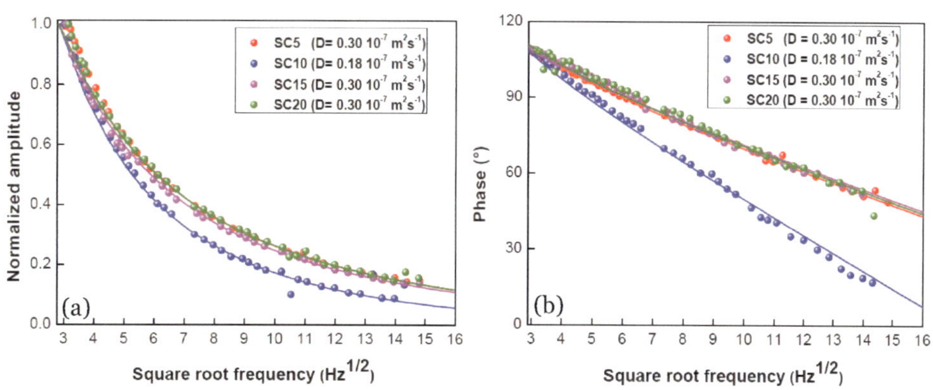

Figure 7.
The normalized amplitude (a) and phase (b) of experimental photothermal signal versus square root of modulation frequency of the magnesium oxychloride cement with PVAc fitted with theoretical curves (line).

photothermal signal versus square root of modulation frequency of the magnesium oxychloride cement with PVAc fitted with theoretical curves. The theoretical curves that best coincide with the experimental curves allow deducing the thermal diffusivity of the samples (**Figure 7**). The difference between theses curves is due to their different thermal diffusivity. We note that the addition of PVAc significantly influences on the diffusivity values. Indeed, the thermal diffusivity decreases with the percentage of PVAc and reaches their minimum values at 10% and begins to increase after this value. The reduction of thermal diffusivity of cement is due to the insulating effect of polyvinyl acetate particles. The thermal diffusivity of PVAc is measured by the same technique (PTD) [12].

5. Determination of the thermal conductivity of the samples (with black layer)

In order to determine the thermal conductivity value of the sample, we have deposed a thin ink layer of few microns thick at the sample surface which we have taken into account in our theoretical model. The ink layer absorbs the entire light beam and therefore considerably increases the amplitude of the photothermal signal and makes the signal sensitive to the thermal conductivity of the sample.

5.1 Thermal conductivity of Sorel cement without PVAc

Indeed, these curves represent the experimental and theoretical variation of normalized amplitude and phase of the photothermal signal versus the square root of the modulation frequency (**Figure 8**). The best adjustment is obtained for a thermal conductivity equal to 0.9 $Wm^{-1} K^{-1}$. This suggests that the thermal conductivity (0.9 $Wm^{-1} K^{-1}$) and thermal diffusivity (0.4×10^{-7} m^2 /s) values are reasonably accurate.

5.2 Thermal conductivity of Sorel cement with PVAc

The curves on **Figure 9** show the experimental and theoretical variation of the normalized amplitude (a) and phase (b) of the photothermal signal with the square root of modulation frequency. The best adjustment of experimental and theoretical curve leads to the determination of thermal conductivity value with precision. We notice that magnesium oxychloride cement (without PVAc) present the greatest

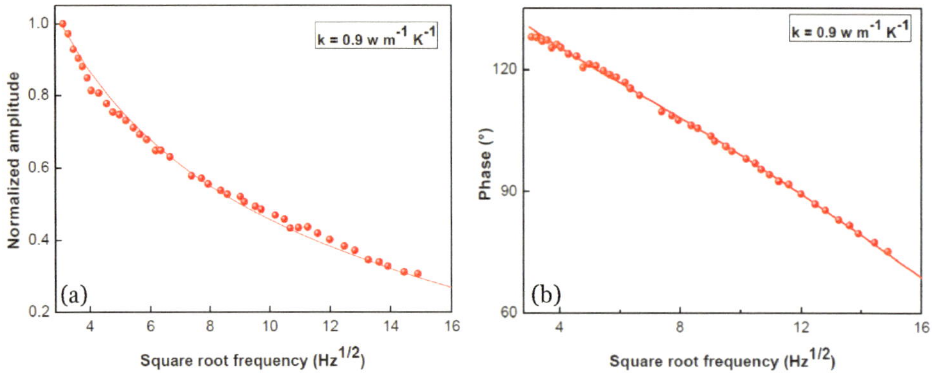

Figure 8.
Normalized amplitude (a) and phase (b) of experimental Photothermal signal versus square root modulation frequency of PVAc fitted with theoretical curves (line).

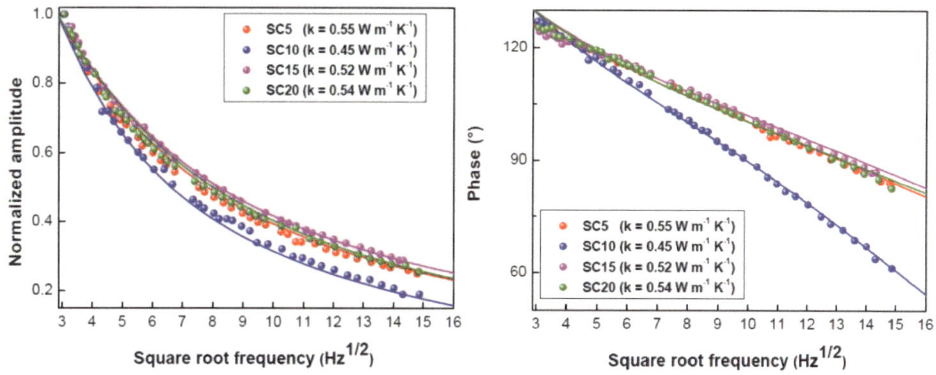

Figure 9.
Experimental and theoretical variation of the normalized amplitude (a) and phase (b) of the photothermal signal with the square root of modulation frequency.

values of thermal conductivity. The samples of magnesium oxychloride cement with different percentages present a minimum of thermal conductivity at 10% addition with a slight increase to 15 and 20% of PVAc. Indeed, the insulating effect and the amorphous structure of PVAc particles [11–12] are responsible for the reduction of thermal conductivity. We conclude that the incorporation of PVAc improves the thermal properties of the MOC. Values of thermal conductivity and thermal diffusivity decreased from 0.9 to 0.45 w/mk, and 0.4×10^{-7} to 0.18×10^{-7} m^2/s, respectively. Moreover, an optimum of 10% is obtained, which corresponds to a reduction of approximately 50% in thermal conductivity and 55% in thermal diffusivity which can be explained by the appearance of the smallest pore-size in the cement matrix [12].

6. Compressive strength

In order to study the effect of the incorporation of polymers on the mechanical properties, compressive strength measurements made on samples air-cured for 28 days. The results of compressive strength are shown in **Table 1**.

The incorporation of PVAc losses the mechanical strength and it presents a variation from 64.88 to 23.07 MPa corresponds to a reduction approximately of 64%. Furthermore, the reduction in compressive strength of the thermal optimum is approximately 55%. However, it maintains good mechanical strength. Indeed, we improve the thermal properties of magnesium oxychloride cement while keeping a good resistance. The phase 5 ($5\,Mg(OH)_2.MgCl_2.8H_2O$) is the source for strength and hardening of Sorel cement. At the atomic scale, phase 5 formation can be explained by the adsorption of the atoms Mg^{2+}, OH^- and Cl^- when mixing MgO and $MgCl_2.6H_2O$ on the surface of MgO. **Figure 10** shows the microstructures of both magnesium

Samples	Compressive strength (MPa)
SC0	64.88
SC5	41.33
SC10	29.05
SC15	26.40
SC20	23.07

Table 1.
Compressive strength of Sorel cement for different percentage of PVAc.

Figure 10.
Microstructure of SC5, SC10, SC15 and SC20.

oxychloride cement without and with PVAc (5 μm). We see in these images the needle shaped crystals. These needles shaped crystals is the P5 ($5Mg(OH)_2.MgCl_2.8H_2O$). We note that the incorporation of PVAc does not affect for the development of P5.

7. Conclusion

The objective of this work is to study the thermal properties of Sorel cement, which is a building material with promising performance. In order to determine the thermal properties of the cement, the photothermal deflection technique is used. The insulation properties of the SC were enhanced with the incorporation of PVAc. The thermal conductivities and thermal diffusivity of SC with PVAc are in a range of 0.45–0.9 W/m K and 0.4 10^{-7}–0.18 10^{-7} m^2/s, respectively. A thermal optimum is obtained for 10% of PVAc so that a reduction of 50% of the thermal conductivity and 55% of the thermal diffusivity is obtained. Furthermore, the compressive strength of SC is notably influenced by the incorporation of PVAc and increases with decreasing PVAc percentages, the value of compressive strength vary from 64.88 to 23.07 MPa and achieves a reduction of up to 64% for 20% of PVAc. However, the thermal optimum reaches 55% reduction in compressive strength (29.05 MPa) but retains good mechanical strength.

Acknowledgements

This chapter was supported by a grant from Ministry of Higher Scientific Research of Tunisia. The authors would like to thank the technical team of the physics mathematics quantum modeling and mechanical design laboratory in Preparatory Institute for Engineering Studies of Nabeul and the National Research Centre of Materials Sciences in Borj Cedria Technological Park.

Author details

Rim Zgueb[1*], Amal Brichni[2] and Noureddine Yacoubi[1]

1 Physics Mathematics Quantum Modeling and Mechanical Design, Preparatory Institute for Engineering Studies of Nabeul, Nabeul, Tunisia

2 Useful Material Valorization Laboratory, National Center for Research in Materials Sciences, Technopole Borj Cedria, Soliman, Tunisia

*Address all correspondence to: rim.zgueb@fsb.u-carthage.tn

References

[1] Sorel. Improved composition to be used as a cement and as a plastic material for molding various articles. U.S. Patent; 1866. pp. 53-92

[2] Saad SM, Nasser S, Aboterika HA. Removal of synthetic reactive dyes from textile wastewater by Sorel's cement. Journal of Hazardous Materials. 2009; **162**:994-999

[3] Maravelaki KP, Moraitou G. Sorel's cement mortars decay susceptibility and effect on pentelic marble. Cement and Concrete Research. 1999;**29**:1929-1935

[4] Ji YS. Step by step building of a model for the berkovich indentation cycle. Materials Letters. 2002;**56**:353-356

[5] Cole WF, Demediuk H. X-ray thermal and dehydration studies on magnesium oxychloride. Division of Building Research. 1954;**5**:234-251

[6] Alegret S, Blanco M, Subirats R. Potentiometric study of the reactivity of calcined magnesium for use in magnesium oxychloride cements. Journal of the American Ceramic Society. 1984;**67**:579-582

[7] Beaudoin JJ, Ramachandran VS. Strengh development in magnesium. Cement and Concrete Research. 1975;**5**: 617-630

[8] Yu H. Magnesium Oxychloride Cement and Its Application. 1st ed. Beijing: China Building Materials Industry Press; 1993

[9] Zongjini L, Chau CK. Influence of molar ratios on properties of magnesium oxychloride cement. Cement and Concrete Research. 2007;**37**:866-870

[10] Fernelius NC. Extension of the Rosencwaig-Gersho photoacoustic spectroscopy theory to include effects of a sample coating: Applied Optics. 1980: 23-16

[11] Zgueb R. Contribution to the Study of the Thermal Properties of Sorel Cement by the Photothermal Deflection Technique. Tunisia: University of Carthage; 2019

[12] Zgueb R, Brichni A, Yacoubi N. Improvement of the thermal properties of Sorel cements by polyvinyl acetate: Consequences on physical and mechanical properties. Energy and Buildings. 2018;**169**:1-8